TRIGGER

GUARD

A WRITER'S GUIDE TO FIREARMS

TRIGGER

GUARD

A WRITER'S GUIDE TO FIREARMS

CHRIS GRALL

TACTIQUILL INC

For more information, or to book an event, contact:

Chris@ChrisGrall.com

http://www.ChrisGrall.com

Book design by Chris Grall

Cover art by Ravi Zupa
MT-SMG-U6 ©RaviZupa 2015
RaviZupa.com

Cover design by Chris Grall and Nathaniel Marunas

Illustrations by Chris Grall and Wiki Commons

Edited by: Kim Ledgerwood and Kimberly Hunt of Revision Division

ISBN – Hardback : 979-8-9901343-3-1

ISBN - Paperback: 979-8-9901343-0-0

ISBN - eBook: 979-8-9901343-1-7

ISBN – Audio Book: 979-8-9901343-2-4

Contents

List of Images

The following is a list of images in this book, not created by the author.

The authors of these images, along with the websites they are hosted on, in no way endorse the subject matter or content of this work.

All effort has been made to fully attribute these images and comply with the terms of their licensing agreements.

E-11 Blaster Rifle, Author: JMC, Page 22 – Public Domain
https://commons.wikimedia.org/wiki/File:StormTrooper_Blaster.jpg

Gun Powder Recipe, Author: PerioclesofAthens, Page 30 – Public Domain
https://commons.wikimedia.org/wiki/File:Chinese_Gunpowder_Formula.JPG

Hand Cannon, Author: Viollet-le-Duc, Page 30 – Public Domain
File:Dictionnaire raisonné du mobilier français de l'époque carlovingienne à la Renaissance, tome 6 - 357.png - Wikimedia Commons

Arquebus (cropped image), Author: Viollet-le-Duc, Page 30 – Public Domain
File:Dictionnaire raisonné du mobilier français de l'époque carlovingienne à la Renaissance, tome 6 - 359.png - Wikimedia Commons

Bandolier, Author: Rijksmuseum, Page 36 – CC0 1.0 Universal Public Domain
https://commons.wikimedia.org/wiki/File:Bandelier_met_kruitmaatjes_en_kogelzakje,_NG-1149.jpg

Alexander Forsyth c. 1820, Author: Unknown, Page 62 - Public Domain
https://commons.wikimedia.org/wiki/File:Alexander_Forsyth_c._1820.jpg

Gatling Gun, Artist: Matthew Trump, Page 73 – This file is licensed under the Creative Commons Attribution-Share Alike 3.0 Unported license.
https://commons.wikimedia.org/wiki/File:Gatling_gun.jpg

Minigun, Artist: Paul Hagerty, SSgt, USAF, Page 73 – Public Domain
https://commons.wikimedia.org/wiki/File:HH-3-minigun-vietnam-19681710.jpg

Dedication

This guide is dedicated to all you writers out there.

Way back when I was in the Army, I had a book with me wherever I went. In that hurry-up-and-wait world, you need something to take your mind off the tedium. For me, it was books.

When I was in Special Forces Selection, there would be events during the day, and we had downtime, but we were only allowed to sleep four hours a night. We needed a way to fill the void, and books were the answer. They became a valuable trading commodity. No lie, I traded books to make trades for other books.

I brokered deals between my fellow candidates to hunt down a valuable title. You see, we could only bring one book with us, and the selection process lasted three weeks. Guys who washed out would leave their food and their books behind as a symbol of brotherhood and support for those of us who remained.

On deployments, there was always a free library available. Take a book, share a book. Guys always abandoned their stash at the end ... except that one for the ride home.

This work is dedicated to the authors of the past who helped me through tough times.

Of course, the kids today have it much easier than I did back then. With the advent of electronic readers and tablets, all you need is an internet connection, and you can get whatever you want.

So, this book is also dedicated to you, the writers of today and tomorrow who keep the next generation entertained through the boring times.

Thank you,
Chris

Disclaimer

You know I've got to do the legal thing at some point; let's knock it out now and get it over with.

This book is meant to be a guide for authors who write guns into their stories. It will cover some tactics and techniques, but it is not a training manual. People who wish to receive firearms training should seek out a qualified instructor.

This book discusses how firearms operate. It is not a technical manual. Any reference to how a specific firearm operates should be confirmed by reading the instruction manual for that specific gun before actually handling or firing it.

The author is not responsible for damage or injury, intentional or unintentional, to any person or property as a result of the information presented in this guide. This work is solely for writers of fiction.

The author does not support or condone illegal or unlawful activity. Gun laws differ between jurisdictions. Always check applicable local, state, territory, and federal laws concerning the sale, purchase, transfer of ownership, possession, carrying, storing, or manufacture of firearms before doing the same.

Safety Brief

Before we get started, I must give the obligatory safety brief. This is common sense advice for handling any firearm. If your character is a law enforcement officer or member of the military, they've heard this a million times. Gun ranges may have a more in-depth list of dos and don'ts. These rules are the minimum and should be followed by anyone handling a firearm.

However, your characters can break these rules all they want.
- Treat every firearm as if it were loaded.
- Never point a firearm at something you aren't willing to kill or destroy.
- Keep your finger off the trigger and outside the trigger guard until you're ready to shoot.
- Be sure of your target and what's behind it.

You are sure to see the big four, or some version thereof on any gun range. Some range rules include:
- Keep your gun on safe until ready to fire.

This is good advice, but as you will see, it only applies to certain firearms as a best practice.

I'll add two of my own:
- Never hand anyone a loaded weapon.
- When you are handed a weapon, immediately check to see if it's unloaded.

The last step may seem redundant, but better to be redundant than have a firearms accident. With firearms, you rarely get a do-over.

Foreword

When I received a mysterious email from a stranger professing to know many things about guns, little did I know that shot across the bow would turn into not only a massive benefit for my writing career, but also result in a treasured friendship.

First, a quick bit of backstory.

In my early 20s, I set a goal of becoming a full-time fiction writer. Whatever it took, that's what I'd do. I spent the next twelve years writing, submitting, going to conventions to meet editors and publishers, researching the business, hunting for an agent, and all the other things new writers do to find a way into the game. The thing that finally cracked the code for me was self-recording my own audiobooks and releasing them as serialized, weekly podcasts. After doing four full-length novels in this fashion, I built up a great fan base. When I released my novel Ancestor as an indie-press trade paperback, that audience showed up in spades. The book hit #1 on Amazon's Horror and SciFi charts, and was the #2 book overall on Amazon, second only to Harry Potter and the Half-Blood Prince. This is before e-Books, mind you – it was print vs print, me against all books big and small.

While that heady height lasted only a couple of days, New York Publishing saw this unknown book from an unknown writer with no marketing budget whooping the tar out of titles with multi-million-dollar ad campaigns. In short, Big Publishing lost its little mind. My novel Infected went to auction, I got a wonderful deal from Random House, and boom, goal achieved – I have been a full-time author ever since.

Enter that mysterious email…

Chris Grall was a fan of the podcast. During those first four books I mention above, I got delightful (and often quite strange) emails from fans all over the world, including many altruistic communiques from subject-matter experts who politely offered their help in making my fiction more accurate. Doctors, lawyers, physicists, biologists, oh my, people kindly told me that while they loved my stories, I was getting a lot of real-world facts wrong. That old adage of "write what you know?" I knew little about alien invasions, cops tracking serial-killing monsters, brain science, paleontology, and — you guessed it — firearms. One of those emails was from Chris, who informed me (in a very classy way, I might add) that I knew diddly squat about guns, and it showed in my work.

Cliff Notes version: he was correct. I'd fired a .22 once when I was all of eight years old, and that was the sum total of my firearms experience. Oh, and also from 80s cop shows and action movies, where guns were always portrayed accurately…

Chris offered to be my resource for all things firearms. I don't think he had any idea of what he was getting into, because I started emailing him question after question, which led to phone conversations about what gun to use in what scenario, which eventually led to him reading entire drafts of mine so he could find mistakes on things I had no idea were wrong, and one amazing trip where I helped him train SWAT teams and got shot in the face several times.

In my endless pestering of this smart, awesome, helpful military veteran, Chris started to build up a knowledge base of what writers really need to know to accurately portray firearms in their stories. Why is this body of knowledge critical? Because if you get guns wrong, many readers will notice, and it takes them out of the story. Even if you lose willing suspension of disbelief for just a moment or two, it diminishes your work and shortchanges your readers. When your heroine shoots at the bad guy, and her semi-automatic pistol clicks on empty three times before she throws it at him? In a literary sense, you have just shown your ass, and readers know you didn't bother to do even the tiniest bit of research on how things actually work. You have disrespected the veterans, police, hunters, and general firearm enthusiasts that gave their precious time and (hopefully) their hard-earned money to consume your work.

Why get it wrong and disrespect your audience when you can read this book and get it right?

That's what Trigger Guard does — it helps you get it right.

As a novelist himself, Chris reveals the info you need that will show your respect for your readers. Trigger Guard will make you sound like you know what you're talking about, whether you've never seen a Glock in real life, or you have a collection of seventeen shotguns.

If you've read this far, you may have already bought the book. If so, dig in.

If you're reading this foreword on Amazon and considering if you should purchase this bad boy, consider no more. Buy it, read it, and enjoy the comforting feeling of "gunpowder therapy" without that trip to the range.

– Scott Sigler, *New York Times* bestselling author

Introduction

Firearms are the most frustrating and infuriating things to write into a story. Frustrating, because it can be damn near impossible to find out what you need to know about a certain weapon. I've spent an inordinate amount of time hunting down what the proper terminology is for that one little doohickey on the side of the gun.

Oh, I know what it does, but what is the correct nomenclature? Then, of course, after a half hour of fruitless websites, dead ends, and cat videos, I finally find the term in a Google image search. I return to the manuscript and ... damn it, where was I again?

Infuriating, because even if you get everything right, some knucklehead out there is going to give you a scathing review based on their personal experience or a faulty technique demonstrated in a YouTube video.

The person making this review doesn't realize that your exhaustive research into the policies and procedures for Agency X, during time period Y, was fairly and accurately depicted in your story.

Furthermore, agencies have reasons for the policies they adopt, and trainers have reasons for the techniques they teach. A technique, while sound, may not be acceptable to an agency from a risk standpoint. Some firearm aficionados fail to take this into account.

There's another set of villains lurking in the pages of this guide: gun manufacturers. While researching firearms to use as examples, I uncovered a plot. It seems that, for years, the gun companies have conspired to arbitrarily use different terminology on their weapons for parts that perform similar functions. This is done in an effort to make your life miserable. This guide is an attempt to thwart their evil plans.

Maybe I made that up ... maybe.

And now, my first words of advice and the guiding principle of this book:

Any detail of the firearm, or its operation, that doesn't directly benefit the story, character, or plot should be cut.

Lengthy descriptions of a character's manipulation of the weapon, or descriptions of the firearm, bog down action scenes. Only the gun enthusiasts care about that stuff. By leaving out granular levels of detail, you will avoid most of your problems.

I've designed this guide to be used in two different ways. It is first and foremost a guide for writers. If you flip to The Style Guide section, you'll find my advice and rules for writing in firearms, which will help keep you out of hot water with the firearms community.

This section is cross referenced with the rest of the book and will refer you to places in this text where you can find why the rule is relevant.

The guide is also written to be read straight through. You'll get a truncated history of the development of firearms and learn how the different types function. This culminates in my advice for how to write a gunfight.

Occasionally, I'll be providing examples of common mistakes, how to solve them, or provide background on an issue. These scenarios will provide the context for why a section may be relevant. The format will look like this:

Scenario: Provides context for the error.

Discussion: Where I wax poetic on the intricacies of firearms, how they're used, and why the passage is incorrect.

Possible solutions: Examples of how to get out of, correct, or avoid the mistake.

A Note on Historical Dates.
As you will see, the history of firearms is messy and non-linear. When conducting your own research, you may find dates that conflict with mine. This is normal. Let's take the caplock musket as an example. The percussion cap was patented in 1818. Somebody created or converted a musket to use that firing system shortly afterward. However, the caplock wouldn't be adopted by the US Army until 1842. That's a twenty-four-year stretch. Somewhere in there somebody is bound to say, "Me, me, I did it first." Probably, a lot of somebodies.

In this text I try to note whether the first date of a firearm is design, production, or adoption. But as I said, firearm history is messy.

A Note on Illustrations
All illustrations are my own. I will provide the model of the firearm on which a given illustration is based when applicable, but some illustrations are purely of my own imagination based on historical models.

Many firearms are similar in appearance, and thus, I made no effort to recreate a drawing to suit a minor detail. However, I put a lot of effort into making the illustrations as appealing and informative as possible.

All illustrations are composites of many pictures of the same firearm from different angles with different backgrounds. I alone am responsible for any inaccuracies.

All other pictures or graphics are in the public domain, shared under a Creative Commons license, or my own.

A Note on Gun Laws

This work is purely informative. There will be no discussions regarding the Second Amendment or the NRA. I'm only here to help make sure your work is as accurate as possible.

As I'm sure you're aware, some guns are illegal to own or possess in different jurisdictions. In NYC, for example, you can't even own an airsoft gun without a license. Some jurisdictions allow the possession of a firearm but limit magazine capacity. So, you can have a legal gun, but an illegal magazine. Why does this matter? It doesn't, unless it serves your story.

The legality of the firearm doesn't matter to bad guys. They're breaking the law and/or doing nefarious stuff anyway. What's one more crime? The law is there to restrict access, but again, don't let that bother you.

Bad guys know other bad guys. Criminals know other criminals. Don't let the source of the firearm bog you down unless it's critical to plot.

Gun laws are crazy and inconsistent. Your character can be perfectly legal carrying a pistol with a fifteen-round magazine in one county. However, that same gun or its magazine could be illegal in the next county over.

Final Clarification

There will be places where I say, "Never/do not/don't do this." That advice is for your narration, not your characters. However, if your character is a professional, all the *nevers* and *don'ts* apply. If your character is a novice or inexperienced, they can get as much wrong as you want. Just be sure to let the reader know this is intentional.

Scenario: A firearms novice is getting advice from an expert. The firearm is a Glock 19.

Novice: *So, I put the clip in here?*

Expert: It's a magazine, and yes that's where it goes.

My Potty Mouth

I've been known to use spicy language and drop the occasional F-bomb. I think I caught them all in editing, but don't be surprised if one pops up. Sometimes I get excitable ...

Enjoy!

Firearm Basics

The most basic thing to get right when writing guns into your text is anatomy. You could have your character use an unnamed semi-automatic pistol. However, simply telling me what type of gun is in play implies certain features.

The fact that the gun is a pistol tells me that it probably has only one firing chamber. Semi-automatic tells me that it has a slide and a magazine, which means it probably has a magazine release and a slide release of some type. You can't avoid the fact that there are operational features on a gun, even an unnamed one.

Now, when I say you need to know your gun anatomy, I'm not talking about the buffer retaining pin, the hammer spring, or any other such minor crap. That level of detail is unnecessary unless you're writing a Clancy-esque scene where the failure of some obscure gun part results in an international incident. No, the only thing you need to worry about is what the character interacts with.

This basic anatomy lesson will take care of most of your needs for modern firearms. Just for fun though, we'll use a blaster from *Star Wars* as an example. I know what you're thinking, "C'mon Chris, really, a *Star Wars* gun?" Yup, trust me, all guns have these things: (Notations are my own.)

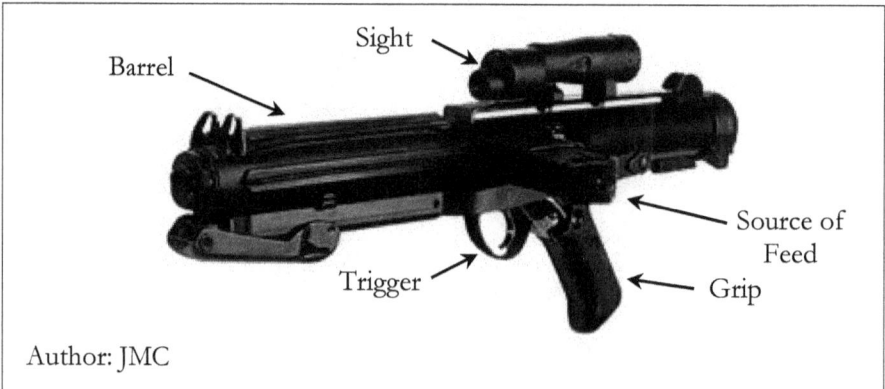

Author: JMC

This example has no visible safety that I could point to and not all guns have a safety the operator can manipulate. However, every gun has a source of feed. That is, something that contains ammunition. In revolvers, it's the cylinder. In automatics, it's the magazine.

Another thing missing from this example is a visible way to manipulate the action. Action being what operates the gun. The Henry rifle has a lever action, the Mossberg 500 shotgun has a pump action, and semi-automatic pistols have a slide. You sci fi folks will want to include a safety and a way to charge the gun ... If that level of detail is necessary.

Back to reality. However, I must first make a quick distinction.

Any firearm that, when fired, cycles its action – extracting the spent casing and loading a new one – without direct work by the operator, is an automatic weapon. This class of firearm is further defined as fully automatic and semi-automatic. Throughout this book, I may refer to the group as a whole as automatics. Pedantic, maybe a little, but I know somebody out there is going to call me out on it. Best to nip that in the bud right now.

Basic Modern Automatic Rifle Anatomy:
Example: M4 Carbine

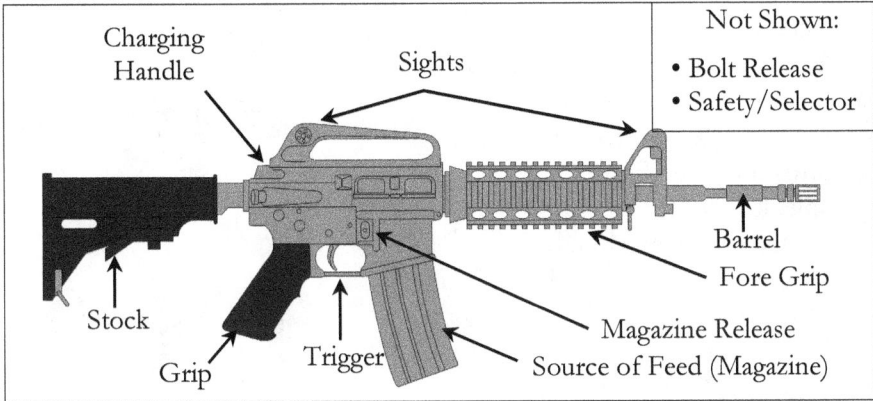

If your character is carrying an automatic rifle, it will have:

- **Barrel:** A tube that is strong enough to withstand the pressure needed to propel a projectile toward its target. In the case of energy weapons, it may contain focusing mechanisms.
- **Stock:** The part of the weapon's assembly used to brace against the shooter's shoulder.
- **Sights:** Mechanical or optical mechanism used by the shooter to aim.
- **Grip/Pistol Grip:** The part of the weapon's assembly that allows the firer to hold it securely.
- **Trigger:** The part of the weapon that initiates the firing of the projectile.
- **Source of Feed:** The part of the weapon that houses ammunition and delivers it to the firing chamber.
- **Safety/Selector:** A mechanism that prevents accidental firing of the weapon or selects a rate of fire.
- **Magazine Release:** A mechanism that allows for detachment of the source of feed for reloading the weapon.
- **Fore Grip:** An area at the front of the weapon, protected from heat, which allows the firer to hold it more securely.

- **Charging Handle:** An external device used to manipulate the bolt.
- **Bolt Release:** A lever or button that releases the bolt from its rear locked position.

BUT! With firearms there's always a but.

Remember that shady cabal I told you about? Those pesky gun manufacturers just love to rename stuff. It's like Ford saying, "I don't want to call it a gas tank. Let's call it the fuel repository instead."

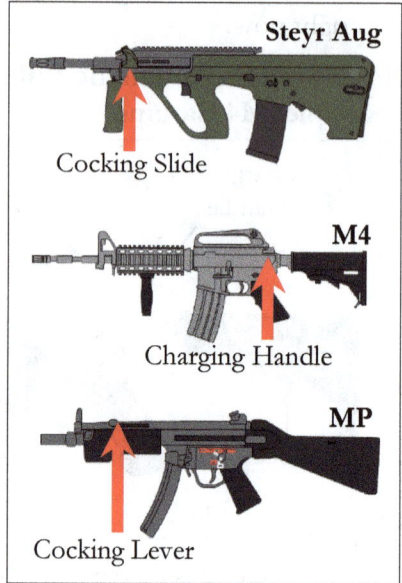

Steyr Aug

Cocking Slide

M4

Charging Handle

MP

Cocking Lever

There are instances where a part is named differently between models or manufacturers even though it performs the same function. As an example, let's look at the device the shooter uses to retract the bolt.

The Steyr AUG calls it the cocking slide. On the M4, it's called a charging handle, and on the MP5, it's called a cocking lever. Different terms for a part that does the same thing: retract the bolt.

If your research doesn't give you the manufacturer's nomenclature, but you know what the thing does, go with the most generic term possible that describes the part's action. Charging handle is the most used term for this device. Let's move on to the pistol before discussing a strategy for dealing with this issue.

Basic Semi-Automatic Pistol Anatomy:
Example: Beretta 92f

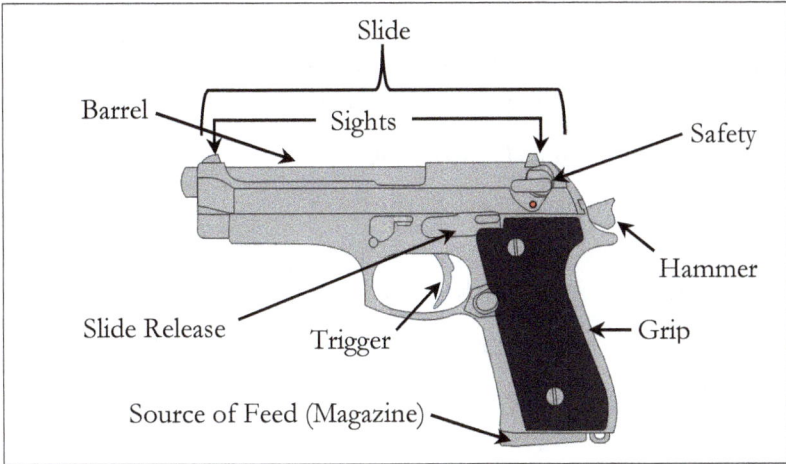

If your character is carrying an automatic pistol, it will have:

- **Barrel**
- **Grip**
- **Trigger**
- **Sights**
- **Safety** (if equipped)
- **Source of Feed**
- **Magazine Release**
- **Slide Release:** Releases the slide when it's locked in the rear position. Similar to a bolt release on a rifle.
- **Slide:** Contains all working parts of the pistol.

Here we have another naming discrepancy. Beretta calls the part a slide release, while Glock calls it a slide stop. Both parts do the same thing, but gun manufacturers can't seem to agree on a standard nomenclature. It's enough to give me fits.

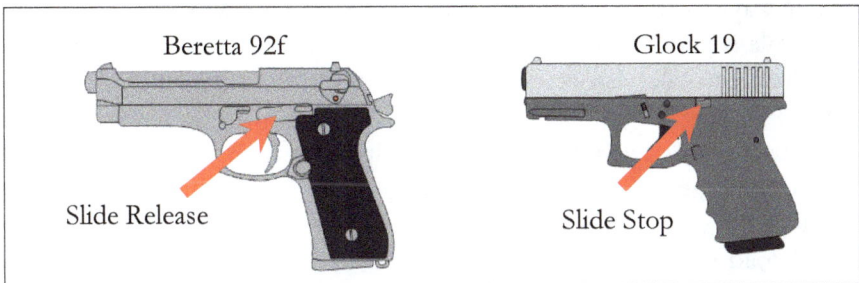

Sure, the *slide stop* on the Glock prevents the slide from going forward, but doesn't it also release the slide when depressed?

Scenario: The writer knows what the part does but can't find the proper term.

Solutions:

Less Detail: This works in almost every case. Let's take the rifle issue first. The doohickey retracts the bolt … and there's the answer.

She retracted the bolt and locked it to the rear.

He worked the action, releasing the bolt.

You don't have to know the name of a thing if you know what it does. The same technique can be applied to the pistol.

He released the slide, chambering a round.

Purpose of Action: The character wants to load or unload the weapon. Hell, I don't really need to give an example but …

She unloaded the rifle.

Writing all the steps involved in a gun procedure can be good for showing character competence, but it takes words. Words better used elsewhere.

She ejected the magazine, retracted the bolt, and inspected the chamber. The weapon was clear. (15 words, no character development, possibility of mistakes in procedure.)

She cleared the gun as she was taught in SWAT school. (11 words, character history, competence established, no possibility for error.)

So, whether the story is a space opera, spy thriller, western, police procedural, or romance you can get around nomenclature problems in creative ways. Keeping character interaction with the weapon as simple as possible is always my go to advice. Here's a list of minimum actions your character will perform with a gun:
- Load the gun.
- Engage/disengage the safety (if equipped).
- Aim the gun.
- Fire the gun.
- Reload the gun.
- Clear the gun (unload).

Examples:

The cop reloaded his pistol.

The soldier looked down her sights at the enemy.

The death trooper put his blaster on safe.

Combining generic actions with undefined firearms can, and often does, work for writers. However, the named weapon adds depth to the story and the character. Let's go western style this time:

He slid the Henry rifle from the scabbard strapped to his saddle.

You've just set boundaries you cannot violate. The Henry rifle is a lever-action rifle that holds fifteen rounds in a tube magazine. You may be able to keep away from the specifics of the Henry rifle and work only in generic terms. However, you do need to know how the Henry rifle operates to ensure that you stay inside the lines. These details may become integral to your plot at some point. Additionally, you need to know the earliest date the Henry rifle was available to your protagonist.

There is no way I can provide essential information for every firearm in this forum. It's just not possible. Hell, not even Jane's *Guns Recognition Guide, Fifth Edition* can do that, and it lists more than 400 guns. One per page. This illustrious volume sits to my left even as I type this.

What I can do is give you a timeline of when firearms of a given type were first produced and how they operate. The next six chapters cover the development and operation of all firearms. No, I will not be covering the oddballs and the evolutionary dead ends. Those are specialty guns that need specific research.

I mean, seriously, if you want to know about the seven-barreled volley gun – 1798 or the harmonica pistol – 1873, you've come to the wrong place. This book deals only with the most common firearms. Which is what you, the writer, should use.

Which is more likely to draw the attention of a gun enthusiast?

She pulled the harmonica gun from her purse ...

She pulled the Colt revolver from her purse ...

She pulled the derringer from her purse ...

All these firearms were available in 1873. Which will be easier for you to research? Which will be easier for you to write?

This brings me to my second rule: Keep it common. You have two rules now:

Rule 1: Keep it simple – less detail is better.

Rule 2: Keep it common – the more familiar the gun is, the better.

Caveat: Unless it serves story, character, or plot.

Sorry, I got distracted for a second. Where was I?

Oh yeah, the next three chapters cover the development of all longarms or shoulder-fired firearms. Interspersed in those pages will be various firearm mistakes covered by the scenario format you've already seen. Fear not, the style guide covers all these issues.

I'll follow the same procedure with handguns. Now, who's ready for a history lesson?

Muzzleloaders

1440–1873

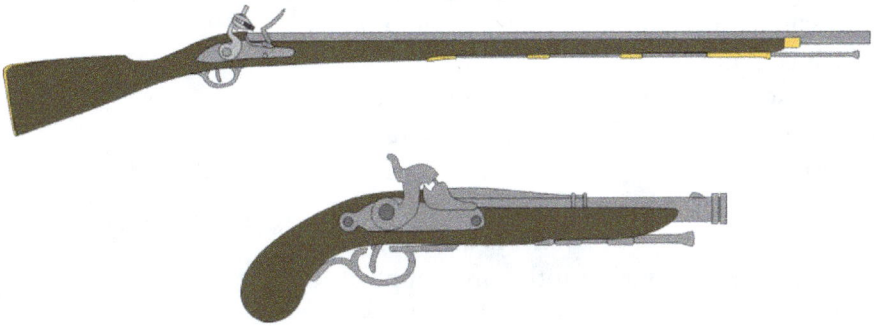

Gunpowder

The first reported formula for gunpowder is credited to a Chinese priest in the year 808 CE. Less than one hundred years later it was used as a tool of war in the form of fire arrows. Next came fire lances and incendiary bombs. Finally, in 1227, CE the first cannon was made in China. By 1280, the formula for gunpowder had made its way to the Middle East and Europe. Cannons were in wide use by 1373.

As metallurgy advanced, so too did the reliability of the cannon. However, big cannons, which could be decisive in battle, were cumbersome and difficult to maneuver. If only they could be carried … Enter, the hand cannon.

The hand cannon was a small cannon tube mounted on the end of a pole. It was held in place by one man while another man applied flame to the touchhole. Around 1400, the slow match removed the need for the second man. Technology marched on and by 1440, the first muskets appeared.

As you may expect, the first firearms weren't very dependable. Military leaders were slow to adopt them because the tried-and-true arrow still had good range and a faster rate of fire. But, their skepticism of the new weapons didn't last long.

In the Battle of Cerignola in 1503, nine thousand French soldiers faced off against roughly six thousand Spanish soldiers. The French relied on heavy cavalry, and the Spanish relied on cannons and the arquebus – an early matchlock design.

It was a disaster for the French, who lost four thousand men. Spanish losses were relatively light at five hundred total casualties. The gun now ruled the battlefield.

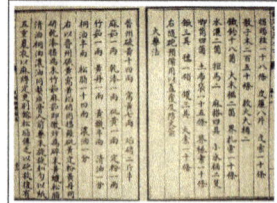

Earliest known written formula for gunpowder, 1044 AD.
- PeroclesofAthens

Hand Cannon
- Viollet-le-Duc

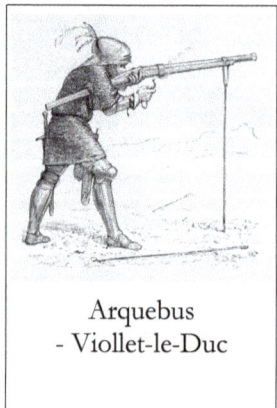

Arquebus
- Viollet-le-Duc

Interlude: The Lock

The lock is the first invention critical to firearm evolution. Before its development, gunners had to apply flame, by hand, directly to the touchhole to fire the gun. This is the basic anatomy all trigger systems are based on. This simplified drawing shows all major components.

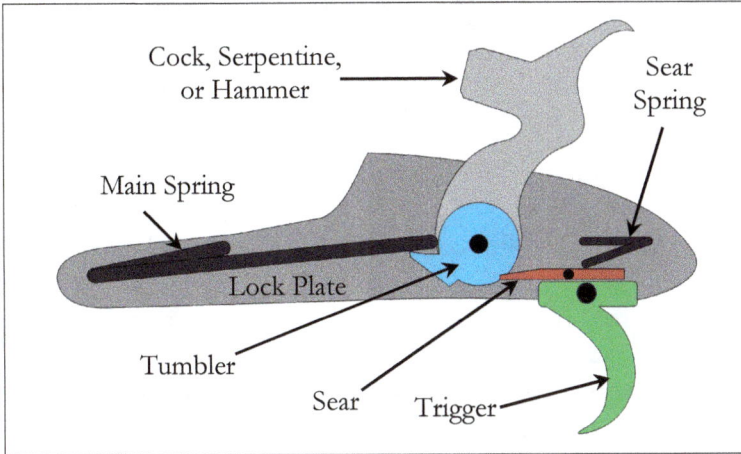

Lock Plate: The platform that holds the assembly together.

Tumbler: Attached to or part of the cock. This component is held in place by the sear and is under tension provided by the main spring.

Main Spring: The spring that drives the cock.

Sear: The component that holds the tumbler in place.

Sear Spring: Provides the force necessary to keep the sear in place to prevent accidental discharge of the firearm.

Cock: Also known as the serpentine, dog, or hammer. This component applies fire, spark, or impact to the priming charge.

Fun Fact: In regard to firearms, cock is a noun, verb, and part of an adjective.

Noun: The part of a lock that holds a slow match or flint

Verb: The action of drawing back a hammer, cock, or serpentine in preparation to discharge a firearm.

Part of Adjective: Describes the position of a hammer or cock indicating readiness to fire.

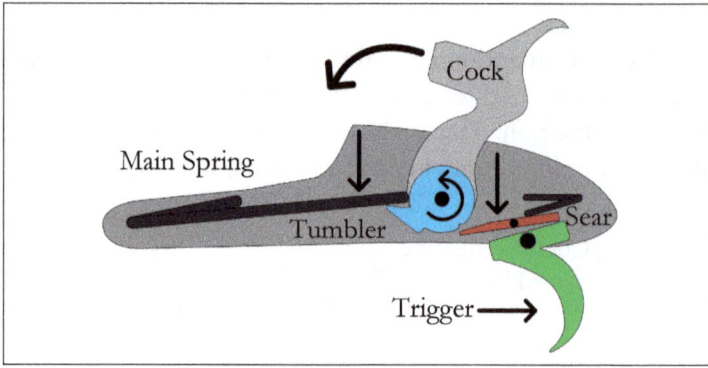

Firing

- When the lock is at full-cock, the whole system is held under spring tension.
- When the trigger is pulled, it applies pressure to the sear, which releases the tumbler.
- The main spring rotates the tumbler, which drives the cock forward, firing the gun.

Half Cock

Matchlock muskets didn't have this safety feature. Later designs of the lock would incorporate a modified tumbler with a secondary position. The half-cock notch was cut in a way that prevented the sear from releasing when the trigger was pulled. This allowed shooters to load

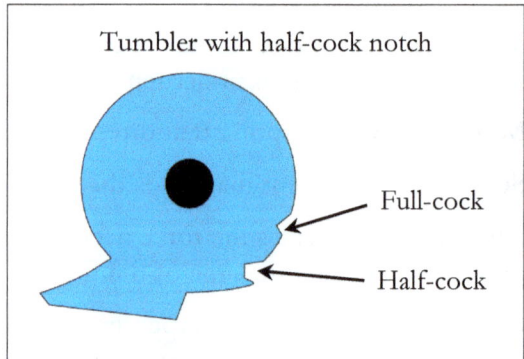

Tumbler with half-cock notch

Full-cock

Half-cock

without fear of accidental discharge. It also allowed the firearm to be carried while loaded.

Fun Fact: Going off half-cocked came from flintlock muskets with faulty locks.

There are many terms used to name early muzzle-loading, hand-fired firearms. I might catch a little heat from the historians out there, but for simplicity's sake, I'm just going to use the term *musket*.

Fun Fact: Firing mechanisms were called locks because they were first made by locksmiths.

Back in the days of the musket, an individual gunsmith produced each gun. There were patterns – archaic blueprints – for firearms, but every gun was handmade. As you'd expect, there was a lot of variation between guns of the same type. For example, the cock could face forward or backward.

It wasn't until the introduction of the flintlock that guns started to take on uniform features. As such, the following illustrations are generic.

Regardless of type, the musket is composed of three main components:

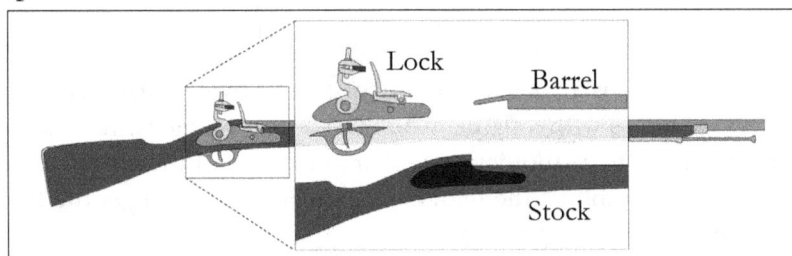

The stock is formed from one piece of wood, and a section was carved out to house the lock.

In addition to the components listed in the interlude, the lock also housed the pan and pan cover.

Barrels were secured to the stock using a combination of screws, bands, or pins, depending on the gunsmith and the overall pattern of the gun. Both muskets and pistols used the same lock systems and loading procedures.

Parts of the musket were often obtained from different sources and could be repaired or replaced individually. The phrase *lock, stock, and barrel* came from having every working part of the musket.

Muzzleloaders are classified by their lock systems, much as firearms today are classified by their action. If the musket had a matchlock, it was a matchlock musket.

It's important to remember that there were no industry standards in this era. New lock designs appeared regularly as gunsmiths sought reliability and safety. I will not be dealing with all the variants. Rule 1: Keep it simple. Instead, I'll cover the four main lock types.

Matchlock, 1440

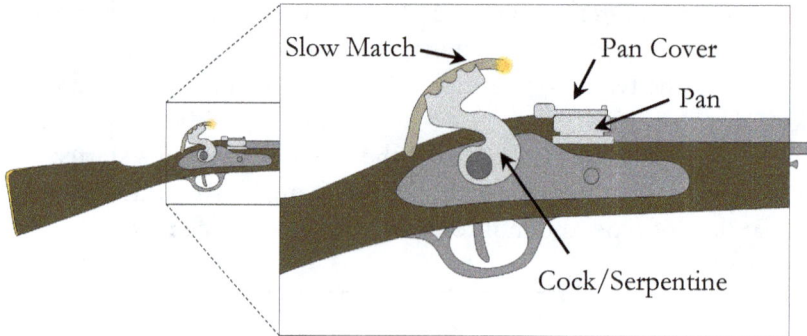

The slow match is a piece of hemp cording or twine. Normally, organic twine burns with a flame. Chemically treating the twine reduced the burning flame to a smoldering ember at the end of the cord, creating the slow match. Combine the match with the lock and you get the first and most dangerous firing mechanism.

Great care had to be taken to ensure that the match was well away from the powder as the gun was loaded. Failure to take this precaution could, and often did, result in calamity.

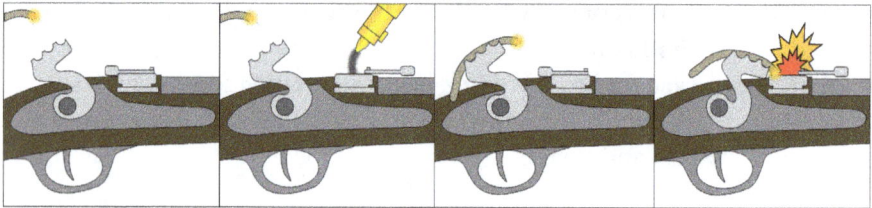

Priming Procedure
- With the musket held level, draw back the cock and remove the match. The match was usually held between the fingers of the hand supporting the musket.
- Open the pan and pour in a small measure of powder. Close the pan and blow clear or shake off excess powder.
- Fit the match back into place ensuring it will fall into the pan when the trigger is pulled.
- Load the musket. There are variations of this procedure. Sometimes the match isn't fit back on the cock until after the loading procedure.
- When the shooter is ready to fire, they open the pan, aim, and pull the trigger.

Video: https://www.youtube.com/watch?v=2KTS8PQ06Qo

Wheellock, 1530

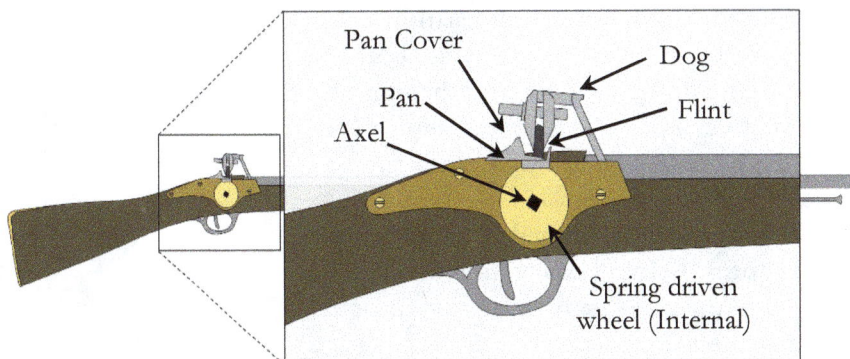

The wheellock represents a big step forward in both reliability and safety. Open flame was no longer necessary to fire the weapon. No doubt, a great relief to everyone. Especially soldiers and their officers.

The cock, now called a dog for this gun, (See, they were doing that naming crap even back then) held a piece of iron pyrite – later replaced by flint, which rested on the wheel.

The shooter used a tool called a spanner to wind the mainspring during the loading process. When the trigger was pulled, it would release the sear and spin the wheel. The iron pyrite would then produce sparks, igniting the charge in the pan, which would fire the gun.

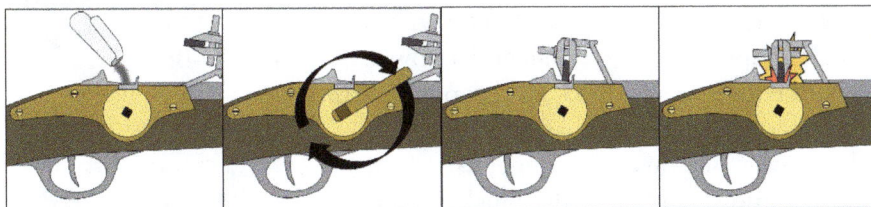

Priming Procedure
- Draw back the dog.
- Open the pan and pour in a small measure of powder. Close the pan and blow clear or shake off excess powder.
- Fit the spanner onto the axel and wind the mainspring.
- Reposition the dog on the pan cover. A spring holds it in place.
- Load the main charge.

The wheellock was a great technological leap forward in lock technology, but it was expensive and overly complex. So, most people used matchlocks until the flintlock was invented.

Video: https://www.youtube.com/watch?v=9YRg2fhy19Q

45

Interlude: The Paper Cartridge, 1580

Until the invention of the paper cartridge, shooters had to measure their powder before loading. Not enough powder, and the ball will either fail to exit the muzzle or fall short of the target. Too much powder could cause damage to the shooter and the musket.

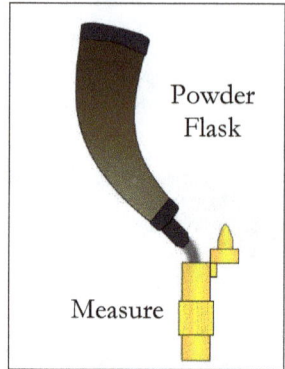

Powder Flask

Measure

Powder was usually carried in flasks or powder horns. Before loading, powder was poured into the measure to ensure the proper amount was used.

Musketeers often carried multiple measures in a bandolier worn crossways across the chest. In this way, they were prepared to fire multiple times without stopping to measure powder.

The well-prepared musketeer also carried their musket balls and pre-cut wadding cloth in a pouch or box.

By the late 1500s, paper cartridges were the standard for most militaries. The paper cartridge combined powder, wadding, and projectile in one easily managed package.

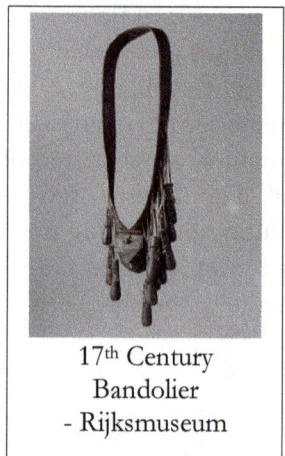

17th Century Bandolier - Rijksmuseum

The cartridge was constructed by wrapping a piece of paper around the ball and a dowel. Once the tube was formed, the proper measure of gunpowder was added, and the end was folded over. As you can imagine, use of the paper cartridge simplified the musket-loading procedure considerably.

Fun Fact: The term *round* came from having enough powder and ball to fire a volley or round. The round encompassed the full procedure of loading and firing the musket.

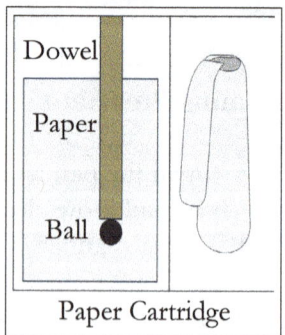

Dowel

Paper

Ball

Paper Cartridge

The paper cartridge contained enough powder and ball for one round. So, the term *cartridge* became synonymous with the term *round*. When the modern cartridge was invented, the term *round* became archaic, but it was already part of the lexicon and remained in use. This is why cartridges are called rounds to this day.

Flintlock, 1630

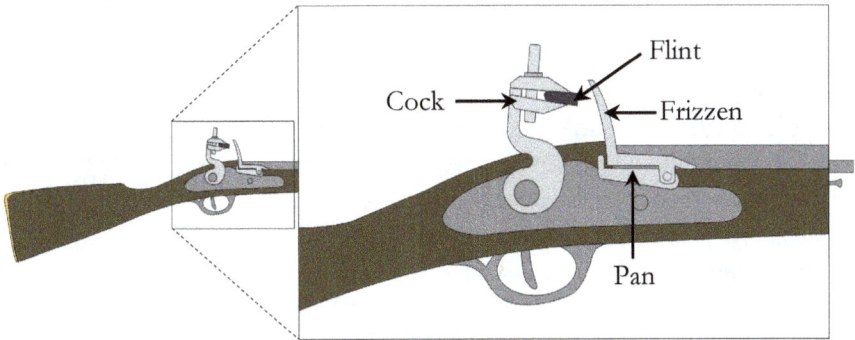

There were many locks that used the flint and frizzen concept, including the snaphance and the snaplock. I'm lumping them all together under the term flintlock because they all operate in the same fashion and they're virtually identical.

When flint strikes steel, it creates sparks. The flintlock design incorporates a steel frizzen for the flint to strike. The frizzen also acts as the pan cover. As the cock swings forward, the flint strikes the frizzen, simultaneously opening the pan and creating sparks.

The flintlock gives us the first firearm with a safety feature: the half cock, as noted in the lock interlude. Half cock is a position of the cock that doesn't allow the gun to be fired. It can be fully loaded and primed, yet still be carried safely.

The flintlock would remain the standard until the advent of chemical ignition.

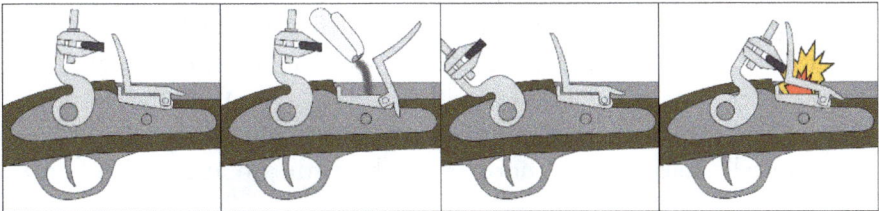

Priming Procedure
- Draw the cock to half cock.
- Open the pan by drawing back the frizzen.
- Add priming powder.
- Close the frizzen.

When ready to fire, the shooter draws the cock back to full cock, aims, and fires.

Interlude: The Rifled Barrel, 1700s

Bowman have long known that a spinning arrow is more accurate than one that doesn't spin. Gunsmiths in the sixteenth century applied this principle to firearms and created the rifled barrel.

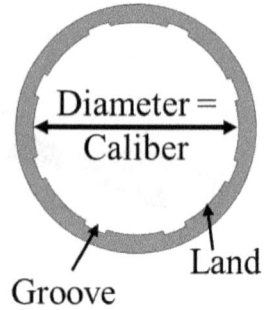

Rifling is achieved by cutting helical grooves down the length of the barrel. Spin is imparted to the projectile as it travels through the barrel.

The raised portions are called lands. The diameter across the lands is what gives us caliber.

Lands and grooves leave their impressions on the projectile. This allows the forensics folks to match a certain fired bullet to a specific gun.

When you research a rifle, you may encounter the term *rate of twist*. This number is expressed as a ratio, like 1:7. That means the projectile performs one full rotation every seven inches.

Note: Rate of twist has nothing to do with barrel length. There's a ton of math behind determining the proper rate of twist for a certain projectile. My brain hurts just thinking about all the variables involved in those equations. Fortunately, somebody has already figured it out. All your character needs to do is buy the right ammo.

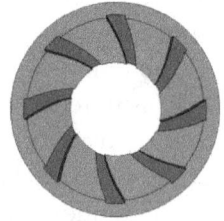

The important thing to know is that if the twist rate is too low, say 1:24, the bullet may tumble. If the twist is too high, 1:3, the projectile may spin too fast and break apart in flight.

The projectile isn't immediately stable when it leaves the barrel. There is a period or distance where the bullet wobbles a bit. With the proper rate of twist, that wobbling goes away and the projectile becomes gyroscopically stable. Some shooters say this is the point at which bullet has "gone to sleep."

So, seventeenth-century gunsmiths knew that a spinning projectile was more accurate than one that didn't spin in flight. The problem wasn't barrel production; they had long since solved that difficulty. The problem was with the ammunition.

The diameter of musket balls in use at the time was smaller than that of the barrel. This meant that, when fired, it bounced around in the barrel before it left the muzzle. To engage the rifling, the fit of the ball had to be snug.

The first solution is to wrap the bullet in an oiled or greased cloth. This filled the void between the ball and the barrel. The American rifle (1700s), also known as the Tennessee, Pennsylvania, or Kentucky rifle, used this solution. It was an incredibly accurate gun used to great effect in early American wars.

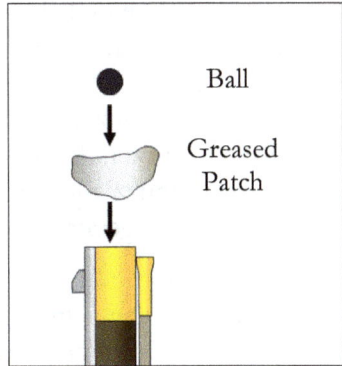

Ball

Greased Patch

"If it was so great, why wasn't it used more than the smooth-bore musket?"

I'm glad you asked. There are several reasons:
- Cost: Rifled barrels are more expensive to produce.
- Speed: Rifled barrels were slower to load than muskets. When you're standing with a bunch of dudes in a line on a battlefield, staring across twenty-five yards at a line of enemy soldiers, firing more often than your enemy is recommended.
- Training: Many conscripts had little to no understanding of ballistics or weapon maintenance. They only needed to know how to stand in a line, load, and fire the musket until the word was given to charge with the bayonet.
- Shot: Cartridges of the day came in a few varieties. One of which was ball and shot, which consisted of one big ball of ammunition and two or three smaller balls. Using shot in a rifled barrel is not conducive to future accuracy … it trashes the barrel.
- Fouling: With heavy use, the rifled barrel tended to build up soot and gunpowder residue. This increased loading time. If left unchecked, fouling could make the gun unusable until cleaned.
- Bayonets: Muskets were made to mount bayonets, and, prior to 1800, rifles weren't. The prospect of a bayonet charge terrified men.

Picture a wild-eyed, screaming man rushing toward you with a two-foot knife on the end of his gun. Not my idea of a good time.

The second solution is to redesign the ball. The Minié ball, 1840s, was a cylindrical, lead projectile with a hollow base. When fired, the base flared out, engaging the rifling. This design removed the need to use a greased patch for wadding, and it cleared up the fouling issue.

However, smooth-bore muskets were still the preferred weapon of war until breech-loading

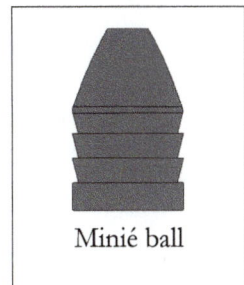

Minié ball

firearms were invented. This is entirely due to the way wars were fought.

So, from the early 1700s forward, we have three variants to worry about:

- Musket: A muzzle-loaded, smooth-bore, shoulder-fired gun.
- Rifled-Musket: A musket with a rifled barrel.
- Rifle: A muzzle-loaded firearm designed as a rifle. Eventually, rifle became the general term for any shoulder-fired gun with a rifled barrel.

If your character is in an army or militia, they will use a musket or rifled musket. The general term *musket* will suffice and, unless added detail is needed, don't use rifle.

If your character is in an army or militia and is part of a sharpshooter or special unit, they will probably use a rifle. Make sure the distinction is clear at the beginning and never use *musket* when referring to that character's gun.

The sharpshooter aimed his Kentucky rifle at the officer in the distance.

If your character is a frontiersman, they will use a rifle. Rifles are hunting weapons.

Wolves, she thought, as she reached for the rifle over the fireplace.

Muskets are weapons of war.

The soldier could load and fire a musket three times in a minute.

Fun Fact: Minutemen, from the Revolutionary War, had to equip themselves. You may be confident in furnishing this character with a rifle. Why purchase a piece of crap musket that's no good for hunting?

However, if the colony is providing the weapon to the Minuteman, it will most likely be a musket. Muskets are cheaper and faster to make.

Caplock (a.k.a. Percussion Lock), 1842 (Adoption by US Army)

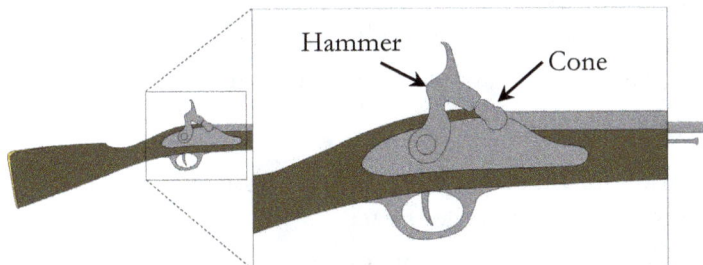

The percussion cap is arguably the most important advancement in firearms technology. The chemicals used in the percussion cap are not important to us, so I won't bore you with them.

What is important, with the cap, is that there is no need for the operator to prime a pan, and the gun could be reliably fired in wet or rainy conditions.

The percussion cap was first developed around 1818, but it took time for armies to adopt them. However, when they did flintlocks were converted by the cartload.

Note, we have another nomenclature change. The cock is now the hammer. Finally, an industry term that sticks.

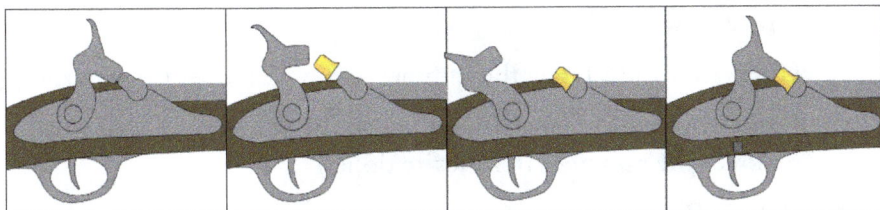

Priming Procedure
- Draw the hammer to half cock.
- Place the cap.

When ready to fire, the shooter draws the hammer back to full cock, aims, and fires. The hammer swings forward, striking the cap, chemistry magic happens, and the gun fires.

Developmentally, caplock firearms were only with us for about twenty years before they were replaced by guns that use the modern cartridge. However, the caplock is still around in the form of modern black powder rifles, which are used during primitive hunting season.

But make no mistake: there's nothing primitive about those firearms and the ammunition they fire, except the fact that they load at the muzzle.

Loading at the Muzzle

I've covered priming in detail, but that is only one of the steps required to make ready your muzzle-loading firearm. The military loading procedure varies depending on weapon type, era, and nationality.

British Flintlock Commands:

1. **Prime and Load:** Hold the musket horizontally, set the cock at half cock, and open the frizzen.

    (Note: The musket is cradled in the crook of the elbow level to the sternum. The lock is centered on the body.)

2. **Handle Your Cartridge:** Remove a cartridge from the cartridge box.

3. **Prime:** Tear open the cartridge using your teeth, and prime the pan.

4. **Shut Pan:** Close the frizzen.

5. **Cast About:** Swing the butt of the rifle around and down. (This gives the soldier access to the muzzle.)

6. **Load:** Dump the gunpowder down the barrel and insert the paper cartridge, with ball.

7. **Draw Rammer:** Draw the rammer and hold it over the muzzle.

8. **Ram Down:** Ram the cartridge down the barrel. (The soldier removes the rammer after they're done.)

9. **Return Rammers:** Restow the rammer in its place.

10. **Shoulder Firelock:** Shoulder the musket, as if ready to march.

11. **Make Ready:** Hold gun in both hands and set the cock to full cock.

12. **Present:** Aim the weapon.

13. **Fire.**

Video: https://www.youtube.com/watch?v=SuYGCji-_5A

Load in Nine Times (Caplock)
(American Civil War)

1. **Load in Nine Times – Load:** The preparatory command, *load in nine times,* lets the soldier know they are about to load the musket as the steps are called. The command *"load"* tells them they are doing it now.

2. **Handle Cartridge:** Remove a cartridge from the cartridge box.

3. **Tear Cartridge:** Tear open the cartridge using your teeth.

4. **Charge Cartridge**: Pour powder into the barrel, discard the paper, and insert the Minié ball.
   – If firing ball and shot, the cartridge paper is put in last to ensure the shot doesn't fall out.

5. **Draw Rammer:** Draw the rammer and hold it over the muzzle.

6. **Ram Cartridge:** Ram the Minié ball or cartridge paper down the barrel. Remove the rammer after you're done.

7. **Return Rammer:** Restow the rammer in its place.

8. **Prime:** Draw the hammer to half cock and place the percussion cap on the cone.

9. **Shoulder Arms:** Take up the musket as if ready to march.

Video: https://www.youtube.com/watch?v=VCAYXQ1Z6q4&t=29s

Note on Loading
The loading procedure can be accomplished in any order. If you want your character to put in the powder and ball first, then prime the pan, there's nothing wrong with that. Especially if they're loading from a powder horn.

I've watched video of a re-enactor loading his flintlock while running. His procedure was powder and ball first, then prime the pan and fire. It was an impressive display.

Video: https://www.youtube.com/watch?v=bflPncephRQ

However, if your character is using paper cartridges, the loading procedure should be pan first, then powder and ball. Unless they have a powder horn handy to perform the priming step.

Any character who is part of a militia or standing army will perform the task as outlined above.

Blunderbuss

What's better than firing than firing one projectile many times? Firing many projectiles at once! I present to you, the blunderbuss. Without which, no list of muzzleloaders is complete.

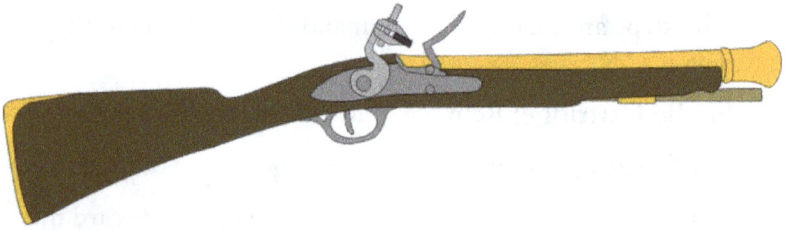

The original shotgun could be equipped with any of the lock systems listed above and also follows the same loading procedures. It's said that the blunderbuss could fire anything from rocks to glass, and even nails. I've no doubt that the myth is true … All you have to do is watch some of the crazy crap people do on YouTube to see the truth in the myth.

However, firing a bunch of oddly shaped metal through your barrel is not conducive to continued accuracy or weapon longevity.

Early models will have the iconically flared muzzle. It was thought that the flare of the muzzle contributed to the spread, or pattern, of the projectiles. This was proven to be false. Pattern, or spread of shot, is determined by barrel length and bore size.

When loading this beast, it's best to use wadding between the powder and the shot, but this wasn't always done. However, using wadding between the powder and the shot does yield a tighter pattern. The firer would definitely load wadding after the shot to keep it from spilling out.

The blunderbuss is a short-range weapon favored by sailors to repel boarders and coachmen to fend off bandits.

The pistol form of the blunderbuss was called a dragon and was an early weapon designed for cavalry. In fact, the term *dragoon* describes a fighting force that is armed with dragons and uses horses for mobility but fights on foot.

The dragon was replaced by the carbine, a shorter-barreled version of a standard musket, but the dragoons got to keep their cool name.

Video: https://www.youtube.com/watch?v=gWbJPI6sNCU

Clearing the Muzzleloader

Modern muzzleloaders are constructed in such a way that the firearm can be easily cleared. There are also modern inventions, like CO_2 cartridges that can blast air through the barrel to remove powder, wadding, and projectiles.

However, these things weren't available to historical shooters. They had only two options. The first method is to simply point the firearm in a safe direction and fire it. The other option was to use a ball puller.

Ball Puller

The ball puller is a device mounted on the end of a ramrod. It was inserted into the barrel and twisted until it either screwed into or captured the projectile. The rod was then removed, taking the projectile with it.

To remove wadding, shooters would use a tool called a worm. This tool could also be used for cleaning. Shooters would wrap cloth around the worm and use it to swab the barrel.

Worm

The tompion is another cool device used in the muzzle-loading era. It was simply a cork or piece of wood stuck in the muzzle to prevent moisture from spoiling the powder.

Soldiers and frontiersmen only needed a few simple tools to maintain their firearms, which they carried in a pouch or cartridge box. Some rifles had hollowed-out compartments in the stock to stow this equipment.

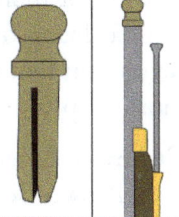

Tompion

Timeline for Muskets

Matchlock – 1440

Wheellock – 1530

Snaphaunce – 1550s

Flintlock – 1630

Percussion cap – 1822

Caplock musket – 1842 (Adopted by British and US armies)

By 1860, the modern cartridge was invented. The Crimean War, 1854-1856 was the last major war in which only muzzleloaders were used. During the American Civil War, both repeating rifles and muzzleloaders were in use.

Other Facts:

Muskets were manufactured by guilds up to the 1700s. This inefficient system often resulted in military units being equipped with muskets with different calibers. A logistical nightmare.

In 1714 the British instituted a manufacturing process based on one pattern for their service musket. Other countries saw the value in standardizing firearm manufacture, but it wouldn't be until the industrial revolution that this concept was fully realized. Thus, a musket is simply a musket up to the point of industrialization.

Common Muskets/Origin/Manufacture Dates:

- Brown Bess – British – 1722–1838 (Also known as the Long Land Pattern)
- Charleville Musket – French – 1717–1840
- Pennsylvania/Kentucky Rifle – American – 1700–1900
- Springfield Models 1795, 1816, 1822 – American – To 1835

Breechloaders

1836–Present

Let's start with a half step forward. The breech is the ideal place to load any firearm. The problem is that the seal between the breech and the barrel must be tight enough to prevent gas leakage. This decreases the energy imparted to the projectile. There are designs for breech-loading firearms that go as far back as the early 1700s.

While some came close to being viable, the most successful was created by John Hall in 1811.

The Hall Rifle, 1819

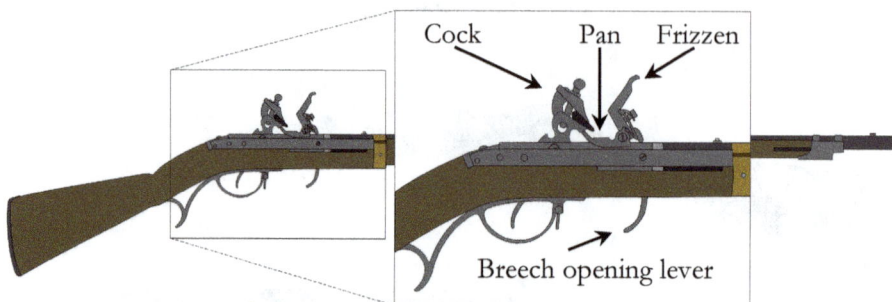

The Hall has a number of innovations not seen in early rifles. The lock is mounted on the top of the rifle. A big hunk of metal on top of the rifle isn't conducive to aiming so the lock is offset to the right and the sights are offset to the left.

Another huge advantage of the Hall is that it's the first rifle ever manufactured with interchangeable parts. This removed the need for a gunsmith if the gun needed repairs.

The loading procedure is similar to the flintlocks of the time.

To Load

- Set the cock to half cock
- Tear the cartridge open with your teeth.
- Prime the pan and close the frizzen.
- Pull back on the breech lever. The whole chamber assembly will tilt up.
- Insert powder, ball, and wadding.
- Close the breech
- Draw the cock back to full cock when ready to fire.

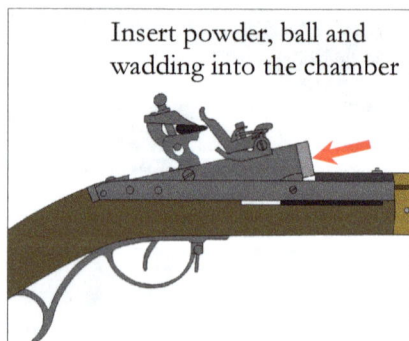

Insert powder, ball and wadding into the chamber

Technically, the Hall isn't really a breech-loader. Nor is it a muzzle-loader. I think of it more as a chamber loader as with some of the revolvers later on in this guide.

The Hall suffered gas leaks at the breech over time. It is a machine after all and parts wear down. However, the design was so successful that when the percussion cap was invented a lot of these guns were converted.

The rifle was produced for the US Army from the 1820s to the 1840s and more than fifty-thousand were produced in three variations. There was the rifle, carbine, and the Hall-North carbine.

This final variation, the Hall-North, is sometimes known as the M1843 or the improved 1840. In this variation the breech lever was moved from underneath the gun to the side.

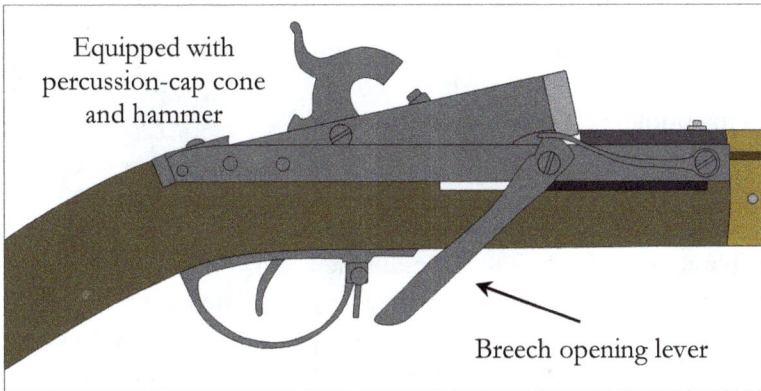

Equipped with percussion-cap cone and hammer

Breech opening lever

The Hall was a technological leap forward both in design and manufacture, but technology advanced during its service and it gave way to the modern cartridge like so many others.

Videos:

History: https://www.youtube.com/watch?v=vpW054cVfHc

Operation: https://www.youtube.com/watch?v=B0IP0Dq3w1o

Interlude: The Modern Cartridge

Life for the writer was so simple once. All you had to worry about was the size of the ball and the type of lock a firearm used. Then the modern cartridge came along and screwed everything up. However, it didn't happen overnight. It took about fifty years. Let's wind back the clock and see how it happened.

It all began around 1805. A guy named Alexander Forsyth was irritated with his flintlock fowling piece – a musket used for hunting birds. The source of the irritation was the fact that, when one pulls the trigger on a flint-lock musket, there is a significant amount of time between when the round is fired and when the projectiles leave the muzzle.

Plus, the sound of the primer charge going off, just before igniting the main charge, tended to spook the target – the birds would fly away. This was unacceptable and Forsyth was determined to do something about it. So, he invented a lock system that used chemicals called fulminates, which would ignite when

Alexander Forsyth

struck with enough force – percussion ignition. Then, he patented it.

As is the case with inventions of this type, people started to copy Forsyth's design; then the patent battles began. The result is the percussion cap. Many people tried to take credit for this invention. All that is known for certain is that the first patent for the percussion cap was awarded to Francois Prelat in 1818.

Now, there are two ways to use percussion ignition. You can redesign the gun and its firing system, or you can make a cap and retrofit existing firearms.

For armies with a large number of flint-lock muskets, conversion is the most cost-effective way to upgrade your firearms. There is the added benefit that the rest of the loading procedure remains unchanged; so, your soldiers don't require much retraining.

Creating a new firing system takes time and experimentation. This is why you'll see innovation in sporting firearms and handguns before they're adopted on a military scale. Consider this: break-action, pump-action, lever-action, and bolt-action rifles/shotguns were all in existence before the American Civil War, yet both the Union and Confederate armies relied on the caplock musket. But I digress.

So, you've got a bunch of evil geniuses running around in the 1800s. All the pieces are in place. They have the science, materials, equipment, and technology to create a fully encased cartridge of ammunition and the guns to fire them. Some of these men were just looking for a better way to kill ducks. Others were out for those lucrative military contracts.

Now we have a chicken and egg problem. To design the gun, inventors had to design the cartridge. To design the cartridge, they had to envision the type of gun they wanted. What followed was an unchecked proliferation of operating mechanisms and ammunition types. There were no standards. They were making it all up on the fly.

The Integrated Paper Cartridge, 1808

The first successful fully integrated cartridge, that is a cartridge that contained the ignition source (cap), powder, and projectile, was designed by Jean Samuel Pauly. His paper cartridge is very similar to the shotgun ammunition of today. Many firearms used integrated paper cartridges.

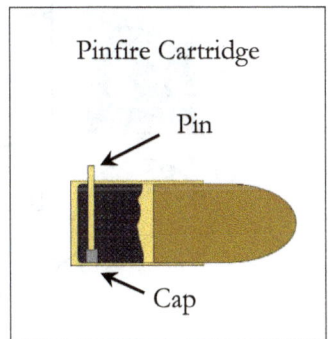

Integrated Paper Cartridge

Needle   Propellant   Projectile

Cap   Buffer

These firearms were called needle guns because the firing needle had to penetrate the cartridge to strike the percussion cap. Advances in materials technology soon made the integrated paper cartridge obsolete.

Pinfire Cartridge, 1832

Casimir Lefaucheux designed a new integrated cartridge that used a pin-firing system. In this design, the firing pin is part of the cartridge. Several guns were designed around this firing system including rifles, shotguns, and pistols.

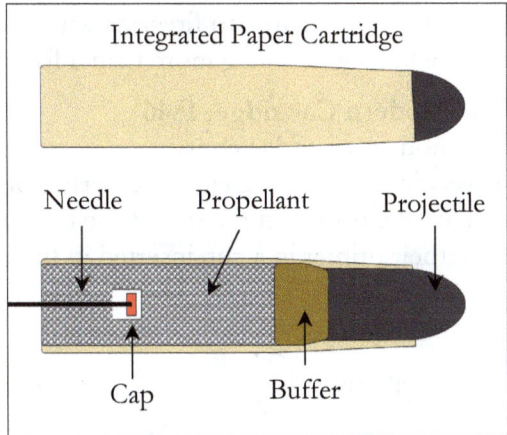

Pinfire Cartridge

Pin

Cap

Finally, Benjamin Houllier, who worked on improving the pinfire cartridge, patented both rimfire and center-fire integrated metallic cartridges in 1846. These two designs were the inspiration for the ammunition we have today.

Before we get into describing the modern cartridge, it's important to note that firearms that used either the integrated paper cartridge or the pinfire system did very well between say, the 1830s and the 1880s.

Needle guns like the Dreyse, (1841) – which were single-shot bolt-action rifles that fired paper cartridges – saw service in combat and over one million of them were produced.

The Sharps rifle, (1849), is another such rifle. This gun featured a falling-block mechanism and was originally designed to fire paper cartridges. The Sharps made the transition from paper to metallic cartridges, but sadly, the company closed in 1881.

The history of these guns is often eclipsed by firearms that had a longer shelf life. The flintlock was the premier firing system for over two hundred years, and the modern cartridge has been with us for almost as long. When compared to firearms with such longevity, it's no wonder they've become nothing more than a footnote.

The Modern Cartridge, 1846

The modern cartridge consists of a casing, primer, propellant, and a projectile. They can be either centerfire or rimfire. Rimfire cartridges contain the primer in the rim of the casing, while centerfire cartridges have their primer in a cup inserted in the center of the casing.

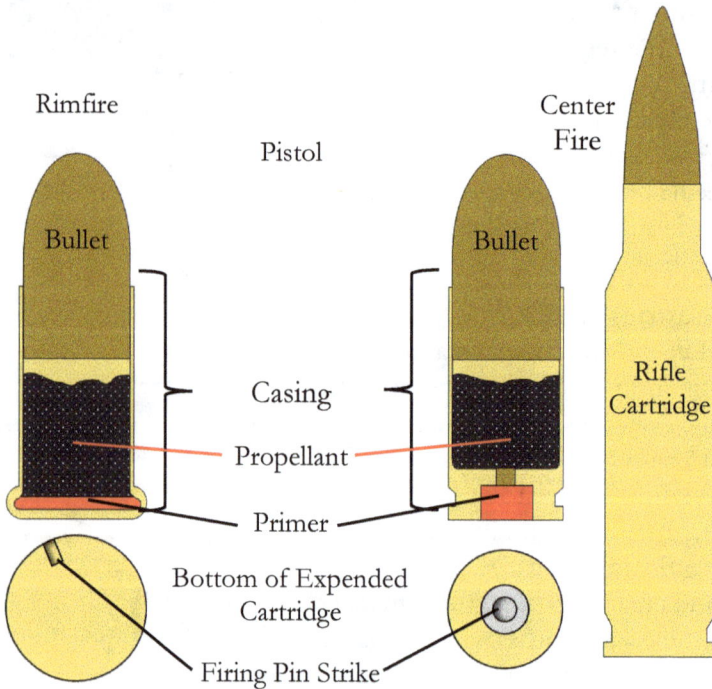

Rimfire

Pistol

Center Fire

Bullet

Bullet

Casing

Rifle Cartridge

Propellant

Primer

Bottom of Expended Cartridge

Firing Pin Strike

A cartridge is also known as a round of ammunition or just a round. Cartridges are sometimes incorrectly called bullets. While incorrect, in the heat of the moment, it is understood that when someone says they need more bullets, they want more ammunition.

As you've probably noticed, there's a big difference between rifle and pistol ammo. Pistols fire shorter, fatter bullets, while rifles fire long, thin projectiles.

Bullet shape and size are directly related to the length of the barrel. With a longer barrel, the bullet is given more time to harness the energy of the propellant. Thus, increasing its velocity.

Rifle bullets, while generally lighter, build better velocity because of their longer time in the barrel, and their shape lends to better penetration.

Pistol bullets don't have the luxury of a lot of time in the barrel and can't build the same level of velocity. They also need to be shorter to reduce the effects of friction. Therefore, their ability to penetrate the target suffers.

To make up for these shortcomings and make the projectile more effective, pistol bullets have more mass. It's all about kinetic energy delivered to the target. If you can't make the bullet go faster, make it heavier so it has more punch.

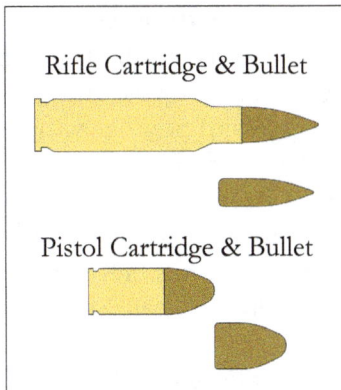

Rifle Cartridge & Bullet

Pistol Cartridge & Bullet

Ballistics: The science of a projectile in flight.

There are three parts to firearm ballistics:

Internal: Travel through the barrel

External: Travel outside the barrel

Terminal: Travel inside the target

Note: Bullet weight is measured in grains (gr). One grain is equal to 1/7,000th of a pound.

Breech-Loading Rifles

When people think of breechloaders, they often picture break-action rifles and shotguns. However, all modern firearms are breechloaders regardless of how the weapon operates.

The invention of the modern cartridge allowed gunsmiths to finally break free from the need to load at the muzzle and breech-loading firearms became the standard. Oh, to be sure, there were breech-loaded rifles and pistols before the invention of the metallic cartridge; they just didn't do very well. Seeing as they're uncommon, they're excluded from this text.

So, it's the mid-1800s, we have a modern cartridge and a butt load of caplock rifles ... What's an army to do? You guessed it, upgrade those puppies.

Now, you'll have to excuse me for a minute because there was a lot going on firearms-wise back then. New designs were popping up all over the place. Firearm development is messy when you approach it chronologically. I'm going to take the evolutionary approach. I'll start with single-shot, manually loaded rifles, then I'll address repeating rifles, and, finally, the automatics. It makes more sense to me this way.

So, armies had a lot of muskets and a way to upgrade them. The idea is simple, chop off the back end of the barrel, creating a breech. Then, give the gun a firing chamber to house the cartridge, and install a hinged breechblock that seals the chamber for firing, but allows for easy reloading. Voilà, you have a breech-loading rifle.

There were many trapdoor designs. There were tilting blocks, rotating blocks, sliding blocks ... the list goes on and on. Nations with standing armies settled on their favorite designs, and the conversion process was put in motion.

In the US, more than 57,000 muskets were converted using the Allin design. Britain upgraded more than 20,000 rifles using the Snider conversion. 5,000 muskets were converted by Irish separatists using the Needham design. And Austria converted more than 120,000 muskets using the Wänzl design.

Note: Some of the upcoming designs were in existence before the date noted next to their type. However, those models aren't illustrated because they used a firing system other than the modern cartridge.

The late 1800s were a politically tumultuous time, and even though there were better options, the armies of the world needed modern guns – and they needed them now. The cheapest way to get them was through conversions.

With so many options to choose from, I'm only going to present you with one as an example. Arguably, it's the "Gun That Won the West."

The Springfield Trapdoor, 1873

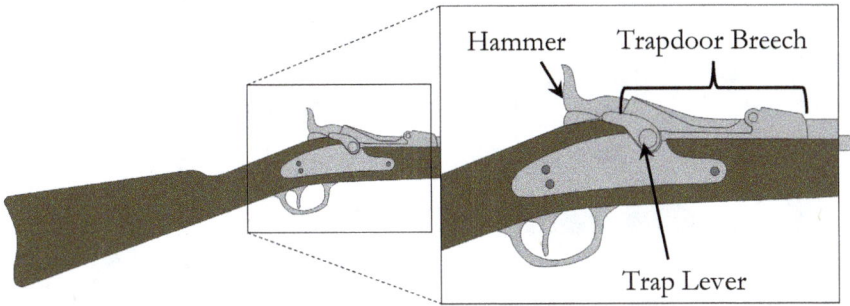

Our only example of a converted musket, The Springfield Trapdoor rifle was the US Army service weapon from 1873 to 1892.

The trigger and hammer of the trapdoor operated similarly to that of the caplock musket.

The safe position was half cock and the hammer had to be drawn to full cock to fire.

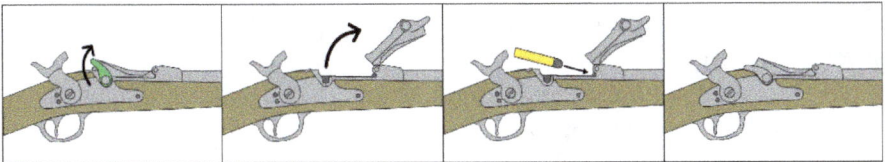

To Load:
- Draw the hammer to half cock.
- Lift up on the trap lever and swing it forward.
- Insert the cartridge.
- Close the trapdoor breechblock. Ensure it is locked back in place.

Note: All firearms from this model forward automatically eject the spent cartridge casing when the action is opened.

Video: https://www.youtube.com/watch?v=JC0C41KX8RY

Interlude: The Receiver

Trapdoor rifles were nothing more than converted, rifled muskets. They used the same basic construction and the same lock and trigger design as the caplock. New firearms were being designed around a whole new concept: the receiver.

Whereas the musket relied on the stock to house all its parts, the receiver sees the stock as just another component.

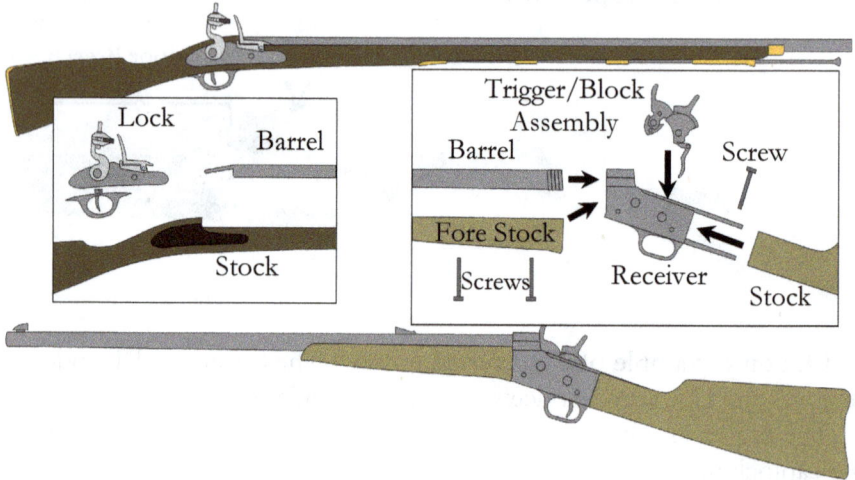

The receiver houses all internal components of the firearm, including, where applicable, the bolt, breechblock, trigger assembly, and hammer. It also allows for the attachment of external components such as the barrel, stock, and fore stock/grip.

Previously, the lock, trigger assembly, and barrel were all mounted individually to the wooden stock. By using receivers, firearm manufacturers could now fully harness the power produced by the modern cartridge.

As you will see with the following single-shot breechloaders, repeating firearms, and the automatics, the receiver allows for precision movement of all parts of a weapon's action.

All modern rifles are based on receiver design, even if it isn't readily apparent.

Rolling-Block Rifles , 1870

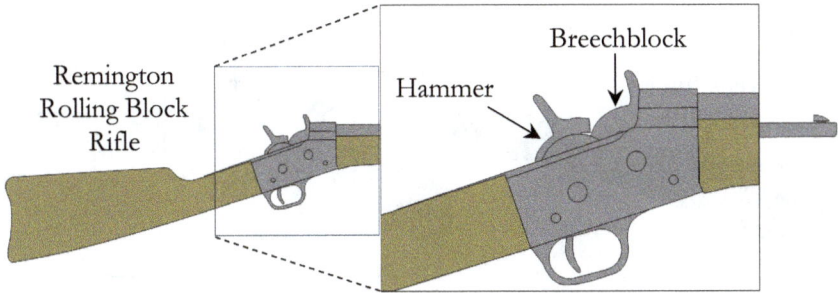

Now that we have the receiver, the breechblock can be incorporated into the weapon itself.

The rolling block is a design that requires the shooter to cock the hammer and manipulate the block during the loading process. It's a step up from the trapdoor systems.

The Remington Rolling Block was an extremely popular gun. Over 1.5 million of these guns were produced between 1866 and 1917.

This firearm was the primary service weapon of several armies.

The safety position for this rifle is half cock. The shooter had to draw the hammer to full cock to fire.

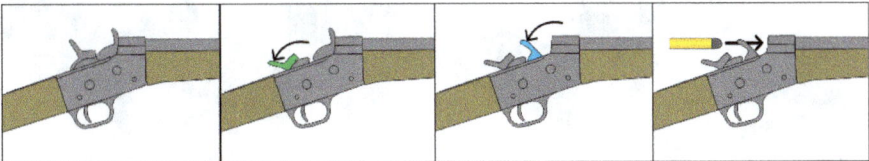

To Load:
- Draw the hammer to half cock.
- Rotate the breechblock backward – exposing the chamber.
- Insert the cartridge.
- Rotate the breechblock back into place.

Video: https://www.youtube.com/watch?v=FNv8hpxDDkU

Falling-Block Rifles, 1878

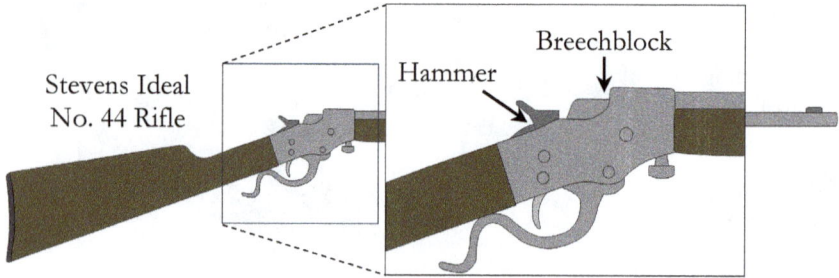

Stevens Ideal
No. 44 Rifle

Hammer   Breechblock

The falling-block rifle uses a lever to operate the block. This rifle can't be called lever action because operating the lever doesn't load a round of ammunition.

This gun is designed to be a sporting rifle used in marksmanship competitions and hunting.

This firearm is fully cocked as the lever is worked down.

Shooters need to manually set the hammer to half cock to be in the safety position.

Modern versions of this rifle have safety levers, buttons, or slides.

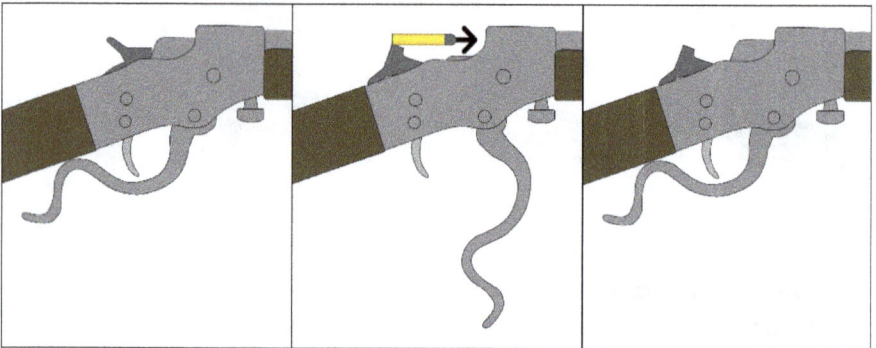

To Load:
- Draw the trigger guard lever all the way down. This drops the breechblock, giving access to the chamber.
- Insert the cartridge.
- Swing the lever back in place.

Video: https://www.youtube.com/watch?v=SJFnDc_m5SY

Break-Action Longarms, 1836

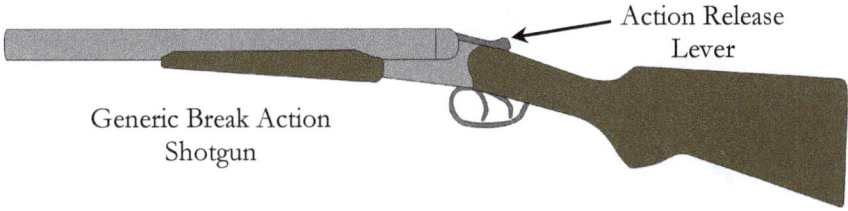

Action Release Lever

Generic Break Action Shotgun

Like all firearms, the break-action variant went through its own evolution. The Lefaucheux gun, 1836, was one of the first. It used pinfire cartridges and the spent casings had to be extracted by hand. (Page 61)

Eventually, designers realized that when the barrels hinged down, they could perform other actions such as ejecting the cartridge and resetting the firing pin. By 1874, the break-action system was perfected.

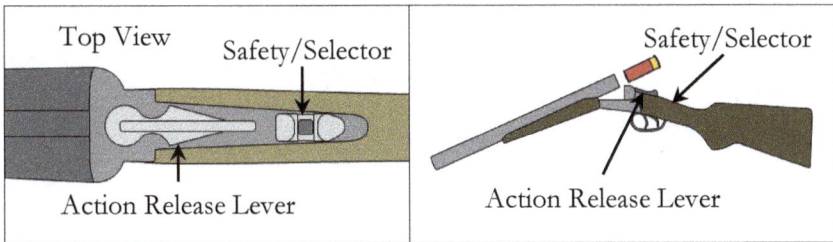

Top View   Safety/Selector

Action Release Lever

Safety/Selector

Action Release Lever

To Load:

- Push the action-release lever to the side. The barrel, or barrels, will hinge downward.
- Insert the cartridge.
- Swing the barrel back into place.

Rifles and shotguns of this type are usually available with one or more barrels.

Double-barrel varieties are of either side-by-side or over-and-under configuration.

The barrel safety switch, or slider, often pulls double duty as a barrel selector. It is usually mounted just behind the release lever.

Older model side-by-sides may have two

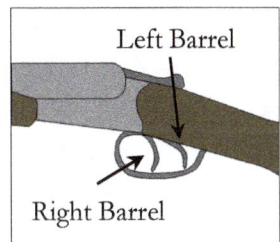

Left Barrel

Right Barrel

triggers. Normally, the front trigger fires the right barrel, and the rear one fires the left.

Video: https://www.youtube.com/watch?v=ZbV3jkgaEek

Repeating Firearms

1860s–Present

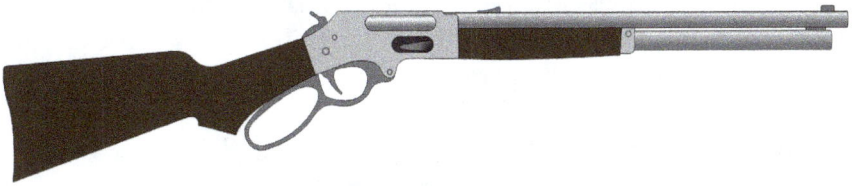

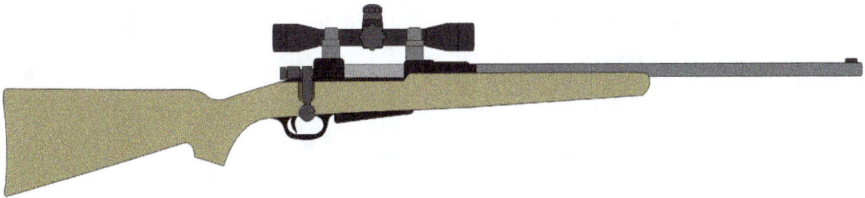

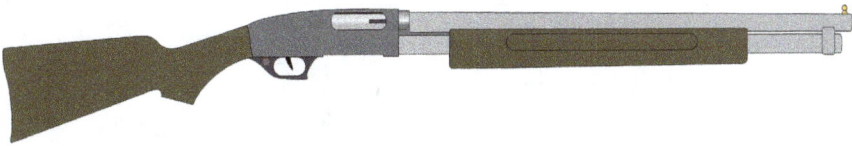

Interlude: The Bolt

The bolt is the next evolutionary step in firearm design and is crucial for the development of repeating guns. In the last three examples, the firing pin was housed in the breechblock, and the ejector, the thing that kicks out the spent cartridge, was part of the receiver. Now, the bolt contains those parts.

The bolt can be operated manually or automatically, and it performs all loading and extraction tasks for the shooter. Rifles and shotguns are now classified by their action. That is, how the bolt is manipulated to do its job.

There are a countless number of bolt designs, and they vary depending on what type of action the weapon uses. This generic diagram gives you the basics of bolt operations.

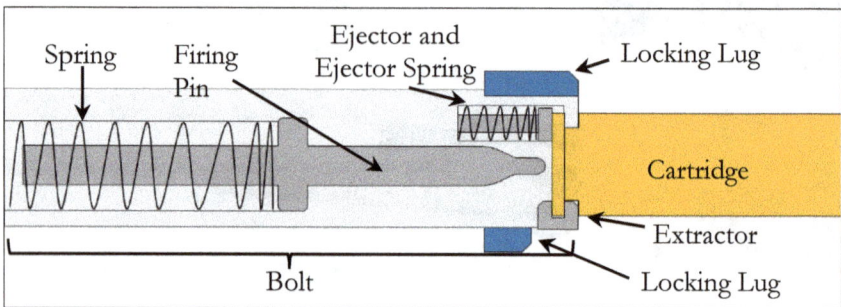

- When the bolt is drawn back, the extractor pulls the expended casing from the chamber.
- Once the casing is clear, the spring-loaded ejector propels it from the gun.
- As the bolt is advanced forward, it picks up the next round from the magazine or a lifter and pushes it into the chamber.
- Some actions cock the hammer or reset the sear on the back stroke, while others perform this action on the forward stroke.
- The locking lugs secure the bolt in the chamber. (Not all bolts use locking lugs.)

Except for the locking lugs, all elements of the bolt are incorporated in the slide of the semi-automatic pistol.

Note: I suggest never, ever, EVER writing about how the bolt operates. This level of detail is full of pitfalls. Stash this knowledge away as useful trivia.

Interlude: The Magazine

The magazine, or mag for short, is the other necessary element for repeating rifles. The magazine stores ammunition until it's ready to be chambered. There are tube magazines, box magazines, and detachable magazines.

Tube Magazines

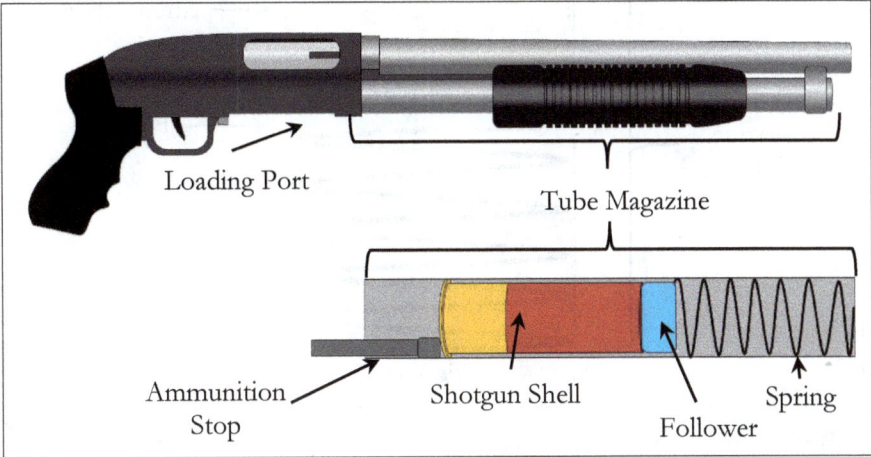

Loading Port

Tube Magazine

Ammunition Stop

Shotgun Shell

Follower

Spring

The tube magazine consists of the tube, spring, follower, and ammunition stop.

When the action is cycled, the ammunition stop is released, and a round is fed onto a loading ramp or lifter. The ramp is raised, and the round is fed into the chamber by the bolt.

Box Magazines

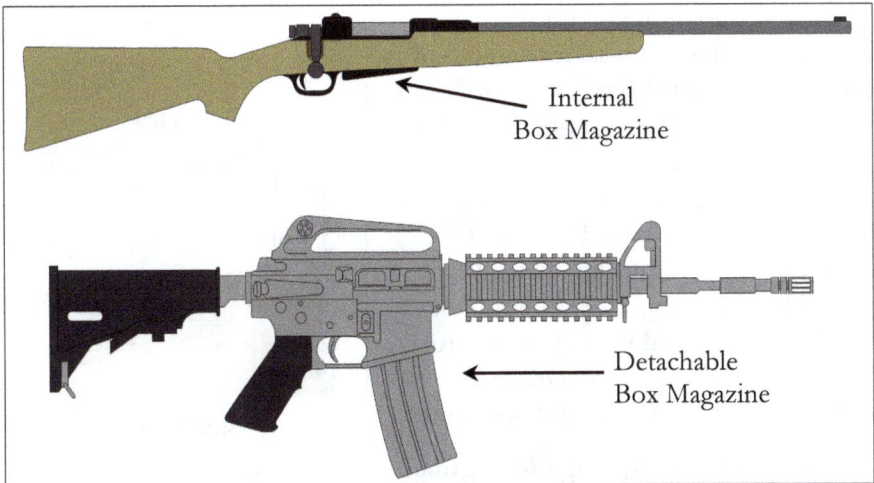

Internal Box Magazine

Detachable Box Magazine

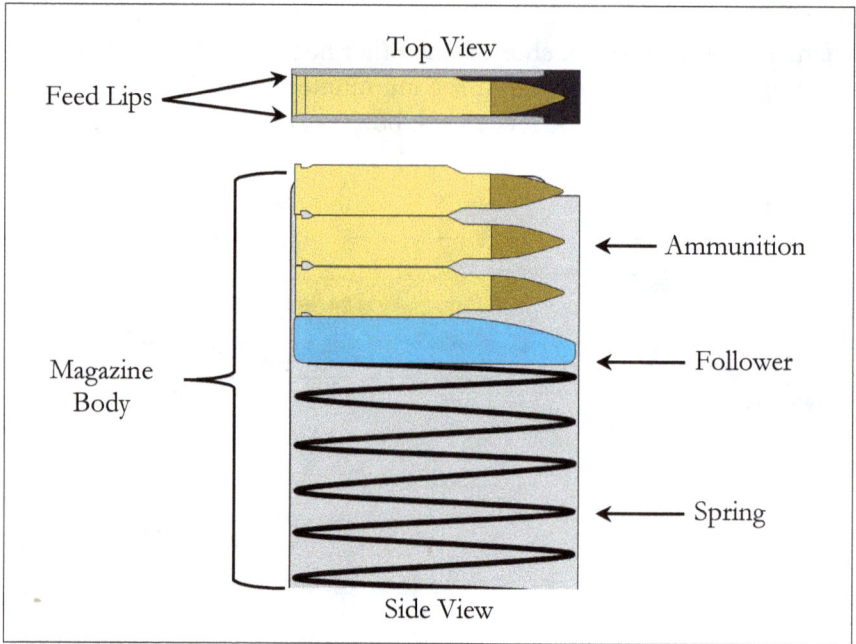

Top View

Feed Lips

Ammunition

Magazine
Body

Follower

Spring

Side View

Both internal and detachable box magazines work in the same way. The magazine is nothing more than a housing for the spring and follower. The feed lips at the top of the magazine keep the rounds in place until they are loaded into the chamber by the bolt, in rifles – or the slide, in pistols.

Most semi-automatic firearms are designed so that forward movement of the bolt or slide is halted by the follower when the magazine is empty. This allows for rapid reloads.

Top-loading magazines can be rapidly loaded by use of a clip. The rim of the cartridges fit into the clip, which is placed into the magazine. The rounds are then slid in, and the clip is discarded.

The easiest way to remember the difference between clips and magazines:

Clips feed mags, mags feed guns.

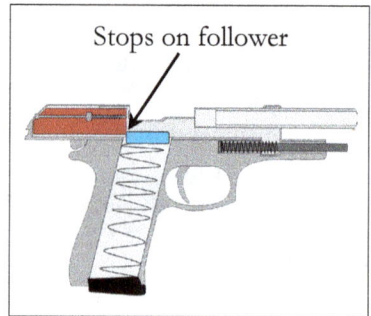

Stops on follower

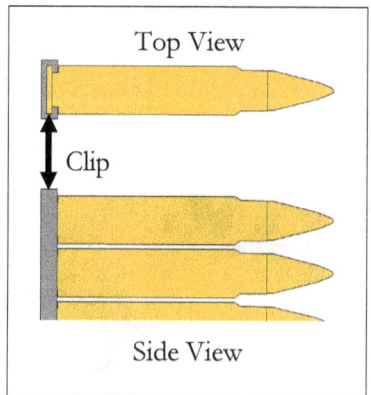

Top View

Clip

Side View

Pump-Action Longarms, 1854

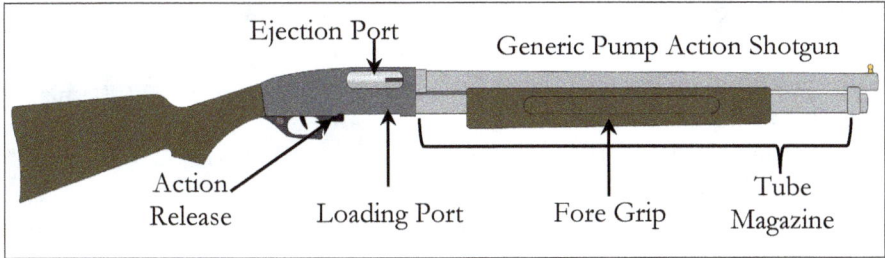

The first slide-action, or pump-action firearm, was patented in 1854 by Alexander Bain. And the first commercially successful pump-action shotgun was the Winchester Model 1897.

When people think about pump-action guns, the venerable 12-gauge shotgun usually comes to mind. However, pump action is also used in some rifles.

Pump-action guns are usually fed by a tube magazine, but some modern versions can use a detachable box magazine.

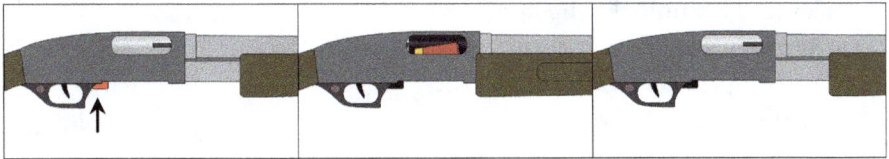

Operation

- With a loaded magazine, press up on the action release lever, which frees the action of the gun.
- Pull the foregrip backward. This will eject any shell in the chamber and release a shell from the magazine onto the lifter.
- Push the foregrip forward. This engages the sear for the firing pin and lifts the shell, which is pushed into the chamber by the bolt.
- When a shell is fired, the action is automatically released, and the shooter can rack in a new round without having to work the action release lever.

Note: Some pump-action shotguns don't have a trigger disconnect. This means that if the trigger is held to the rear after firing, the next shell will fire when the action is cycled.

This is called slam fire, and it was a feature incorporated into shotguns for trench warfare. As a result of accidents, most manufacturers have opted to remove this capability from modern guns.

Safeties

The Winchester 1897 used an exposed hammer to fire the gun. As with most guns of the time, it had no safety. Modern shotguns use the firing pin and sear system. On these guns, the safety is either located on the receiver near the trigger or on top of the grip (tang).

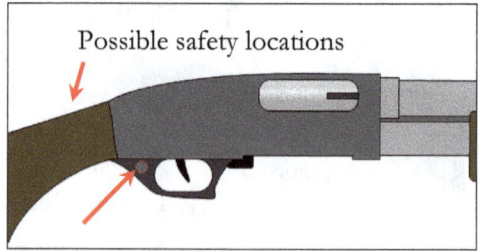

Possible safety locations

To Load
- Feed shotgun shells into the magazine from a port in the bottom of the receiver.

To Unload (Two Methods):
Method One
- Press up on the action release lever. This unlocks the bolt, allowing the shooter to slide the fore grip to the rear, ejecting the shell.
- Cycle the action until the gun is empty.

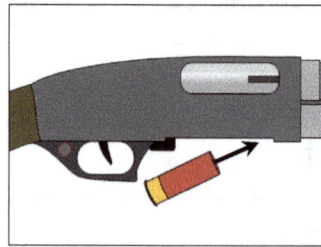

Method Two

If the gun feeds from the bottom, the shooter may have access to the ammo stop. Depressing this lever ejects shells from the magazine without having to work the action. This is safer than cycling the gun a bunch of times.
- Eject all rounds from the magazine.
- Work the action and eject the round in the chamber

Sawed-Off Shotguns

The only reason to saw off the barrel of a shotgun is to make it more concealable; it is no more lethal because it's sawed off. The distribution of shot is a function of muzzle diameter, not barrel length, and the diameter remains the same. Thus, the pattern is also the same.

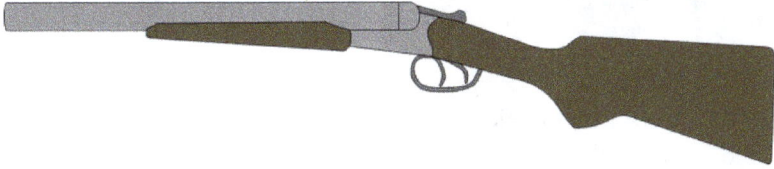

The ATF lists eighteen inches as the minimum legal barrel length for shotguns. So, that's the shortest length they're available in. If your character needs something less than that, I suggest you have them saw off a double-barreled model.

Your character could saw off the barrel of this pump shotgun, but they would likely run into issues with the magazine. The double-barreled model above doesn't have that problem.

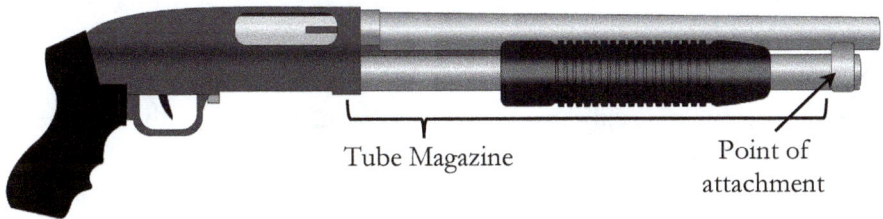

Tube Magazine   Point of attachment

If you're firing slugs through your sawed-off monstrosity, range and accuracy will be decreased. However, if your character is toting around this bad boy, I doubt they care.

Videos:

Shotgun Overview:
https://www.youtube.com/watch?v=21uh28Z77Xg

Slam Fire:
https://www.youtube.com/watch?v=jksldX33HAY

Lever-Action Rifles, 1862

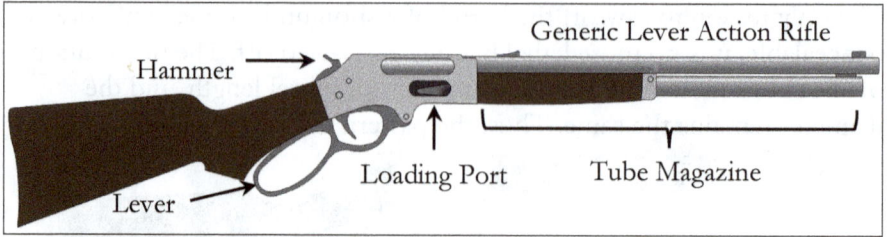

The first lever-action rifles appeared in the mid-1800s and they coexisted with muskets on the battlefield during the US Civil War. Colt produced a couple of designs, but they were never as popular as the Spencer and the Henry rifles.

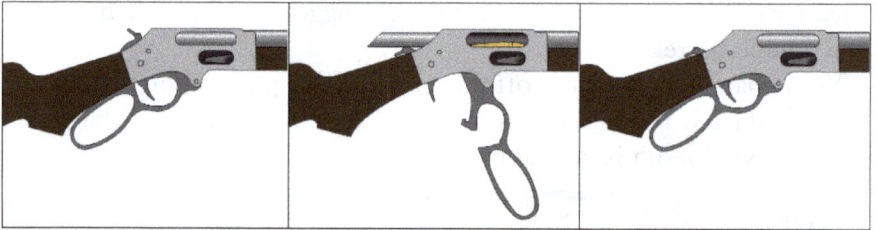

Operation

- With a full magazine, rack the lever down. This cocks the hammer, and a round is then fed from the magazine onto a lifter.
- As the lever is replaced, the lifter rises, the bolt catches the round, and feeds it into the chamber.

Note: Early Spencer rifles didn't cock the hammer as part of the loading process. The shooter had to cock the hammer manually.

Safeties

The first lever-action guns didn't have positive safety mechanisms. Both the Henry and the Spencer rifles did have half cock, and this was considered the safety position of the time. Today, lever-action rifles are equipped with positive mechanical safeties, which will be located on the

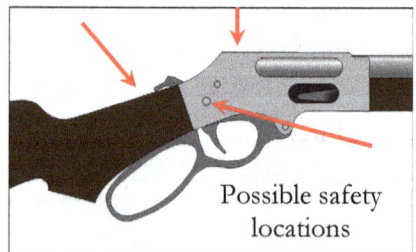

Possible safety locations

top of the gun, the top of the grip (tang), or on the side of the receiver.

Loading

Models like the Spencer had a tube magazine mounted in the buttstock. The shooter would rotate a plate on the butt and remove the spring and follower assembly. They would then insert ammunition, slide the spring and follower tube back into the butt, and lock it in place.

The Spencer had a handy piece of equipment called the Blakeslee box, which held several magazines' worth of ammunition in pre-portioned tubes. This way, the shooter could fully load the magazine simply by dumping a tube's worth of ammo in at one time.

The Henry rifle's magazine was mounted under the barrel, similar to our example. However, unlike the example, the Henry loaded through the front end of the magazine.

In 1866, Nelson King patented the side-loading gate for lever-action rifles, and Winchester would incorporate the loading gate in all subsequent rifles that used tube magazines. These are the lever-action guns you see in westerns. Be warned, if your western takes place before 1866, you should not use a side-loading lever-action gun.

Unloading

The Henry and Spencer rifles are unloaded by removing ammunition the same way it was inserted.

With other gate-loading guns, like the Winchester, simply work the lever action until all rounds have been ejected.

Note: Guns with tube magazines usually fire rounded bullets because rounds are stored end to end. With pointy bullets, there's a risk that the point of one bullet could set off the primer of the next cartridge inside the magazine.

Videos:

Lever Action Overview:
https://www.youtube.com/watch?v=58LbxVd4buo

Henry Rifle:
https://www.youtube.com/watch?v=NbfXjqDzago

Spencer Rifle:
https://www.youtube.com/watch?v=WwhLuhRWYyI

Blakeslee Box:
https://www.youtube.com/watch?v=v4EIB_TiSUk

Bolt-Action Rifles, 1860s

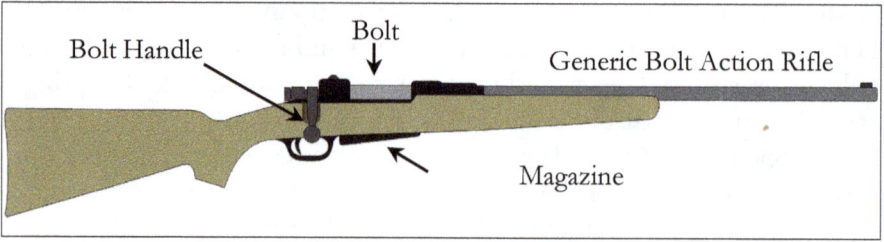

The first bolt-action rifles appeared in the early 1800s. The most famous of these early bolt-action guns is the Dreyse needle gun. It was a single shot rifle that fired the integrated paper cartridge. (Page 61).

The Greene rifle was a single shot bolt-action rifle that coexisted with muskets on the battlefield in the US Civil War. However, the Greene used had a complex loading procedure, which is why it didn't last long before it was replaced. The firing system consisted of a paper cartridge and a separate percussion cap.

The first repeating bolt-action rifles used tubular magazines similar to the lever-action rifles, but they were soon replaced by the box magazine, giving the bolt-action rifle the form it has today.

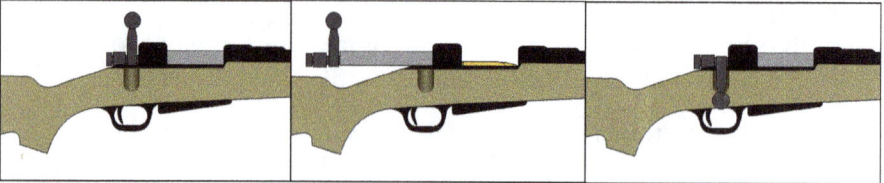

Operation
- With a full magazine, rotate the bolt handle up. This releases the locking lugs that secure the bolt in place when firing.
- Pull the bolt backward. This ejects the cartridge or cartridge case that was in the chamber. It also resets the sear that holds the firing pin in the cocked position.
- Advance the bolt forward. As the bolt travels forward, it picks up the next round in the magazine and loads it into the chamber.
- Rotate the bolt down into place, locking the action closed.

Safeties

Because most bolt-action rifles don't rely on a hammer to fire the round, a positive mechanical safety is needed to guard against accidental firing. Safeties are located on or near the back of the bolt.

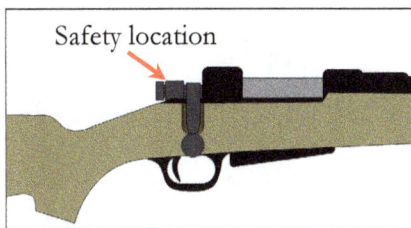

Safety location

Loading

If the bolt-action rifle has an internal magazine, the bolt must be retracted before rounds can be added to the magazine. Some rifles allow for the use of clips to aid the loading process. Simply insert the clip into its slot and press the rounds down into the

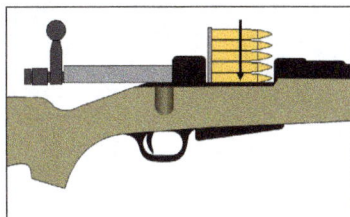

magazine. Otherwise, rounds must be loaded individually.

Unloading

Some variants of the bolt-action rifle have a trapdoor magazine that can be manipulated by a button or lever.

Retract the bolt, ejecting the chambered round, and release the magazine. The stored ammunition will fall clear.

Trapdoor Magazine

If the gun doesn't have this feature, simply work the bolt until all rounds have been ejected.

Some bolt-action rifles have ejecting magazines, and they're inserted and eject similarly to automatic rifles.

Fun Fact: In the early 1900s, bolt-action rifles were chosen over lever-action rifles because they could be easily operated in the prone position.

Videos:

Bolt-Action Overview:
https://www.youtube.com/watch?v=cGbV8hp0pqU

Dreyse:
https://www.youtube.com/watch?v=5qL1g8Hjcpk

Greene Bolt-Action Rifle:
https://www.youtube.com/watch?v=hl4fPfDWiDk

Lee-Enfield:
https://www.youtube.com/watch?v=nSJHmLVAwNI

Springfield 1903:
https://www.youtube.com/watch?v=yJjKH7nPJas

Automatic Rifles
1885-Present

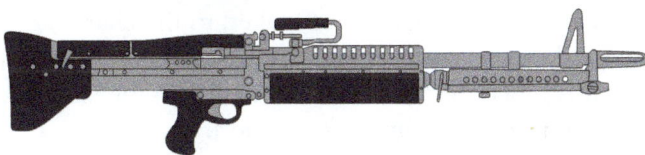

Ferdinand Mannlicher designed and built the first semi-automatic rifle in 1885. Unfortunately, the prototype didn't do very well. Even though the rifle was a failure, the concept and design influenced other firearms engineers around the world. The genie was out of the bottle.

However, there were still some bugs to work out. It wouldn't be until 1937 that a semi-automatic rifle would be adopted as the standard-issue rifle for a national army. That prize goes to the American military and the M1 Garand.

Oddly enough, the Germans had superior machine guns in WWII, yet failed to see the advantage of a semi-automatic infantry rifle. The German infantryman would have to wait until 1943 to have a battle rifle equal to that of the Allies.

Let's look at how automatics function, then we'll continue with a few early models.

Operation

There are four ways to harness the energy from a fired cartridge and use it to operate a firearm. Are they important to you, the writer? No, but I like to be thorough. Here's a quick rundown on how the automatics function.

Blow-Forward Operation

In blow-forward operation, the breach is in a fixed position, and it's the barrel that moves. This form of operation is a total evolutionary dead end. As such, there are few examples of firearms with this operating system.

Examples: Steyr Mannlicher M1894 (pistol), Schwarzlose Model 1908 (pistol).

Blowback Operation

In blowback operation, the bolt is not locked to the chamber and a spring is used to keep the action closed. When the gun is fired, the energy released is transferred through the casing to the bolt face. There are multiple variants of this form of operation.

Examples: Thompson sub-machine gun, Uzi.

Recoil Operation

In recoil operation, the breech and barrel are locked at the moment of firing. They move rearward together for some distance as the bullet travels down the barrel. Once the bullet leaves the barrel, the chamber pressure drops, and the breech opens, ejecting the spent casing. There are three variations of recoil operation: short, long, and inertia. However, that level of specificity is rarely required. Examples: Almost all modern semi-automatic pistols.

Gas Operation

Gas operation harnesses a portion of the high-pressure gas released when the round is fired to work the gun's action. There are two variants to the gas-operated system.

Piston Operation

In piston operation, a portion of the expanding gas that drives the projectile is harnessed by a piston that, in turn, drives the bolt or bolt group.

Long stroke pistons are attached directly to the bolt, while short stroke pistons are separate from the bolt. In short stroke pistons, the bolt is propelled through its cycle by kinetic energy.

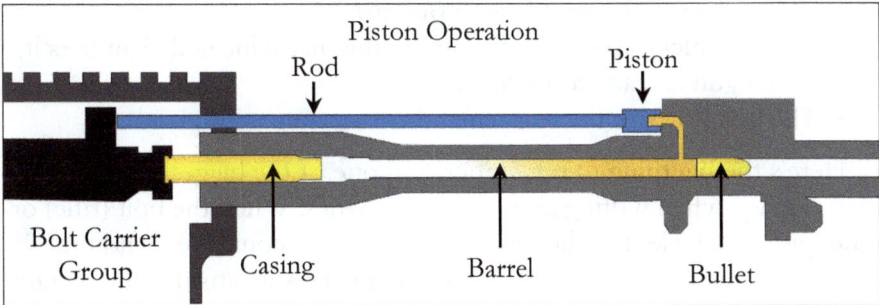

Examples: M1 Garand, FN FAL

Direct Impingement

Direct impingement doesn't use a piston. Instead, the gas travels through a tube and acts directly on the bolt or bolt carrier.

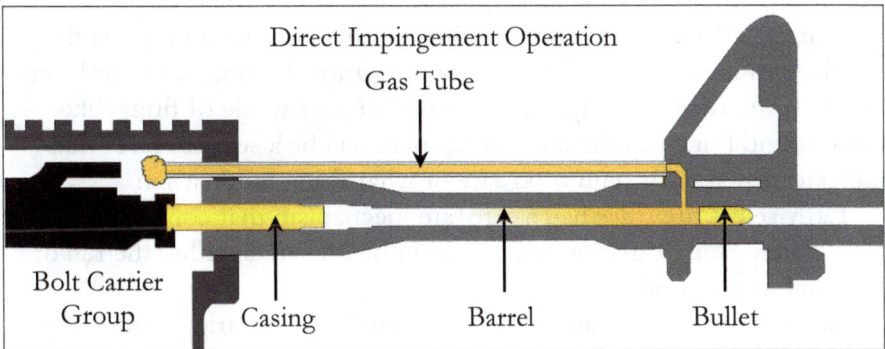

Examples: M16, MAS-49

Operation for Writers

Yeah, yeah, yeah, all that mechanical stuff happens inside the rifle. We don't care about that. What we care about is how it affects the character.

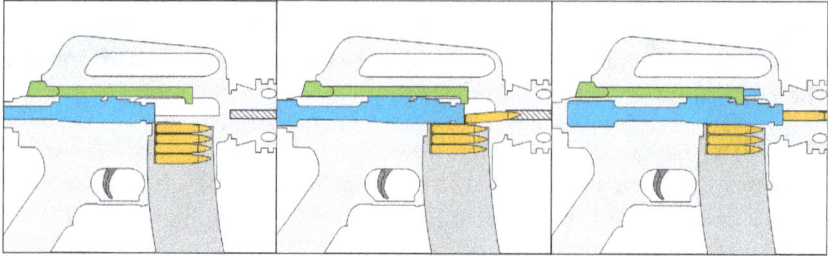

- When the gun fires, the bolt is pushed back by the operating system.
- The spent casing is extracted and ejected by the bolt.
- The bolt re-cocks the hammer or sear.
- The bolt picks up the next round in the magazine and chambers it, so the gun is ready to fire again.
- The bolt locks to the rear when the magazine is empty.

Here's the big thing to remember, and one of the biggest mistakes writers make when writing about the automatics: When the bolt (rifle) or slide (pistol) is locked to the rear, the trigger is disconnected, and the hammer or sear is blocked from functioning. This means the gun will not *click* when it's empty.

Experienced shooters recognize when this happens and, when necessary, perform an emergency reload or transition to a secondary firearm. (See page 114 for more information.)

Safeties

Automatic rifles come in an astonishing array of configurations. If the rifle is semi-automatic only, it may have a simple button, slider, or lever safety. If the firearm is capable of more than one mode of firing, like three-round burst or fully automatic, there will be a selector lever that indicates whether the gun is on safe or what firing mode it's in.

Early select fire rifles had a separate mechanism that controlled the rate of fire. Thus, a rifle or carbine could be on safe in either the full or semi-automatic mode.

Safeties and selector levers will be located near the trigger or on the receiver. There are too many variations to give a simple, single picture example.

M1 Garand, 1937

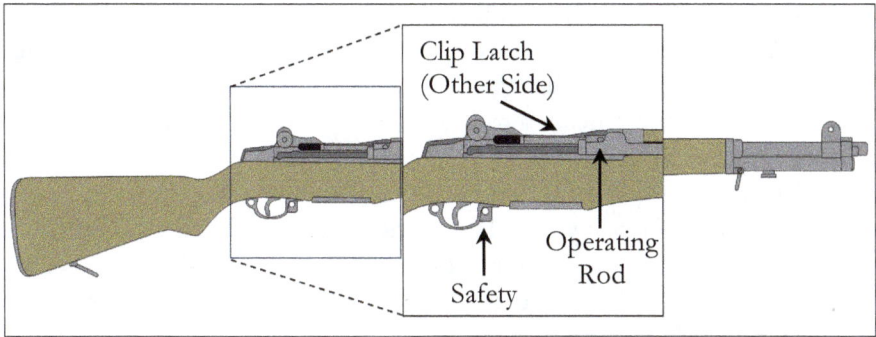

In 1937, the M1 Garand became the first standard-issue semi-automatic rifle for any national military, and it was the service rifle of the US military until it was replaced by the M14 in 1958.

The M1 is a gas-operated semi-automatic rifle fed by an eight-round internal magazine. The magazine is loaded by way of an eight-round en bloc clip. The clip stays in the rifle until the last round is fired, at which point, the bolt will lock to the rear and the clip will eject with an audible ping.

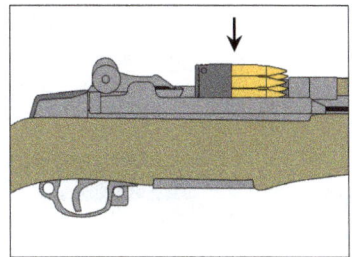

To Load:
- Retract the operating rod until the bolt locks in place.
- While maintaining rearward pressure on the operating rod handle, insert a clip and press it all the way down.
- If the operating rod doesn't release when the clip is fully seated, slap forward on the operating rod.

To Unload:
- Retract the operating rod and hold it to the rear.
- Press the clip latch, and the clip (along with any remaining rounds) will eject.

M1 Thumb
The Garand had a funny little quirk where, if loaded improperly, a shooter could get their thumb smashed by the bolt. If the shooter fails to control the operating rod, when the clip is seated, the bolt is going to go forward with great force. If their thumb is still in the way when that happens … let's just say they won't be happy. It happened so often the affliction was given a name.

FN FAL, 1953

At the end of WWII, as the world settled into an uneasy peace, the North Atlantic Treaty Organization (NATO) was created as a deterrence to the Soviet Union and the Warsaw Pact. It was believed that if NATO had to fight the Soviets, there needed to be, if not a common battle rifle, then at least a common caliber.

NATO was searching for an intermediate caliber. The .30-06 Springfield round, used by the M1 Garand, was considered too big for use in a rifle that fired in the fully automatic mode. Several calibers were tested and discarded. Finally, everyone agreed on a caliber, and the 7.62x51mm NATO round was born.

Firearms companies and national armories were now in a race to create the best design for the new NATO battle rifle. Fabrique Nationale beat everyone to the punch.

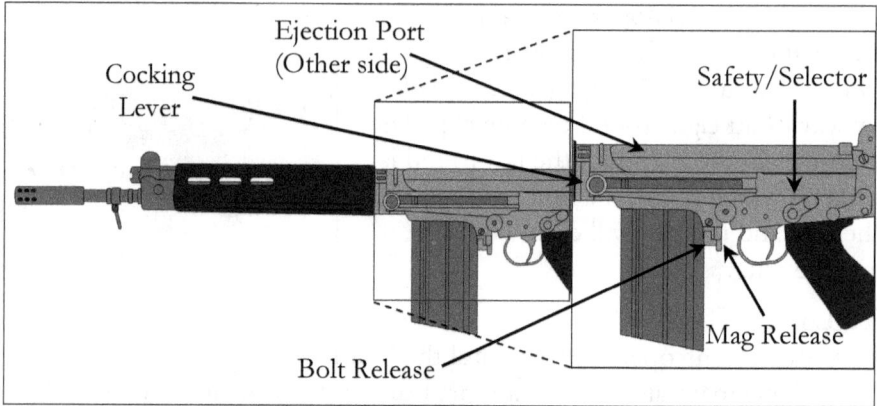

The FAL, fusil automatique léger, or in English, the light automatic rifle, was so popular and so widely used by NATO countries it earned the title, "Right Arm of the Free World." Britain, Canada, West Germany, and over ninety other countries adopted the FAL as their country's battle rifle.

Operation

The FN FAL operates in a manner that most people will recognize. It's fed by an ejecting box magazine and the safety/rate of fire selector is controlled by a single switch in an ergonomic location. The rifle incorporates a bolt release for quick reloads, and expended casings are ejected through a port on the side of the rifle.

The FN FAL is still the issue battle rifle for several countries.

Interlude: Automatic Rifle Loading Procedure

Since we've finally stepped away from top-loading rifles, and we've progressed to modern rifle design, we now have two methods for loading the automatic rifle. One is quick, and the other is more deliberate.

Option 1
- Insert a loaded magazine.
- Tug on the magazine to ensure it's locked in place.
- Retract and release the charging handle.

Option 2
- Retract the charging handle and lock the bolt to the rear.
- Inspect the chamber for debris.
- Insert a loaded magazine.
- Tug on the magazine to ensure it's locked in place.
- Release the bolt.

Why the two options? Because people take shortcuts. Of the two, I was trained to use the second option.

Fully loaded magazines can be difficult to insert properly when the bolt is closed because the magazine spring is already fully compressed.

So, the magazine may not lock in place, but it's up high enough to barely catch on the release mechanism. However, the first round is not high enough to be caught by the bolt. If that happens, when the operator works the action, all they've done is cock the firing mechanism; they haven't chambered a round. This can be very embarrassing.

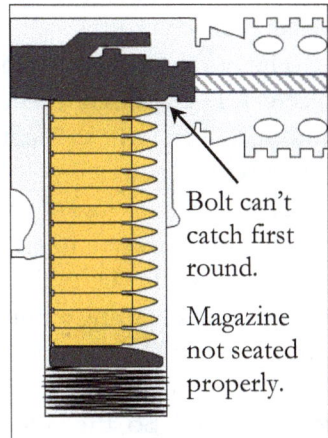

Bolt can't catch first round.

Magazine not seated properly.

I've seen this happen on the firing range.
- The shooter steps up to the firing line.
- The order to fire is given – and click!
- This is soon followed by the magazine falling out. Apparently, the click gives just the right amount force needed to dislodge the magazine.
- Then the pointing and laughing starts.

AK-47, 1947

While NATO struggled to agree on an intermediate caliber for its battle rifle, the Soviets – unhindered by politics and design by committee – had already designed their own new caliber and the rifle to fire it.

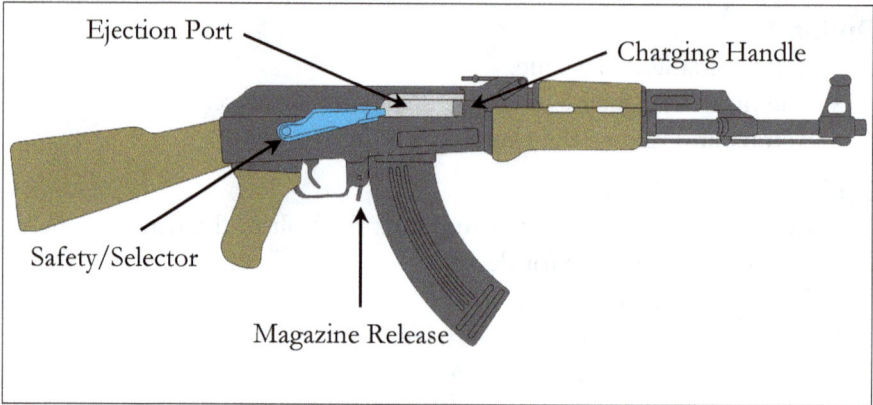

The Avtomat Kalashnikova, selected by the Soviet army in 1947, is one of the most recognized rifles on the planet. Though there have been several iterations of the rifle, any that follow the same basic design are usually called AKs.

The safety on AK variants is an ungainly lever on the right side of the receiver. I've colored it blue for clarity. All the way up is safe. For fully automatic versions, the first click down is full auto, and the second click down is semi-auto.

This seems backward to most people. In western firearms, the selector progresses from safe to semi, then full auto. However, one must consider Soviet-era battle tactics. This rifle was designed to be used by conscripts, at close range, during a human wave assault. For this reason, AK-47s are considered to be less accurate than rifles produced in the west.

The bolt of the AK is not designed to lock to the rear after the last round is fired. So, the gun can click when empty. Loading and reloading will be restricted to Option 1 on page 89.

A shooter can make modifications to the rifle to achieve bolt lock after the final round. If this has been done, the gun won't click on empty.

M14, 1957

As successful as the M1 Garand was, military leaders recognized its drawbacks. As with previous firearms, plans were made to modify and upgrade the M1 rather than replace it. NATO's adoption of the 7.62 caliber changed that. The US almost chose the Belgian FN FAL as the new battle rifle for the military, but they went with an American design instead.

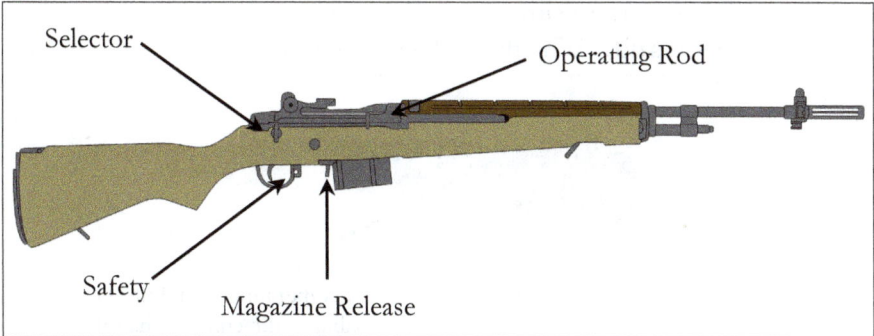

The M14 became the standard service rifle in 1957. However, it was discovered to be too big and heavy for use in the jungles of Vietnam. The M14 was replaced by the M16 less than ten years after it was first adopted. However, it excelled as a marksman's rifle. It's in this role that the M14 is most useful.

The M14 fires the 7.62x51 NATO round and is more effective at longer ranges than the 5.56 cartridge fired by the AR variants. M14s are still in use as designated marksman rifle (DMR). They are used by the Coast Guard for various applications, and some explosive ordinance disposal (EOD) units use the M14 for standoff munitions disruption (SMUD) operations.

Although the M14 is slowly being replaced by more modern rifles, it will probably always have a role in ceremony. The Old Guard, 1st Battalion of the 3rd Infantry Regiment, still issues the M14 as its standard-issue rifle.

The M14 has the distinction as the rifle with the shortest period of service as a standard battle rifle, yet it's also been in service longer than any other rifle in US history.

The M14's bolt locks to the rear after the final round's been fired. So, it won't click on empty. It also has a separate switch for selecting semi- or automatic fire.

Interlude: The Forward Assist and the Reciprocating Charging Handle

The title of this interlude kind of sounds like the title of a children's book. Let's go with it ...

Once upon a time, the US Army wanted a new service rifle. It went to all the manufacturers until it found the one that was just right ... almost.

The Army said to Colt Firearms, "We really like your design, but could you add a way to positively seat the bolt?"

To which Colt said, "Why the hell would you want that?"

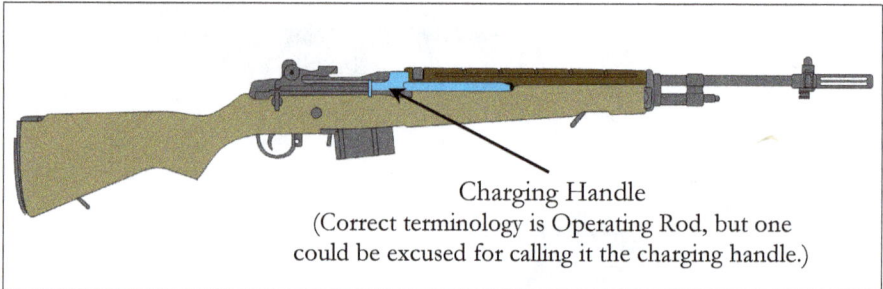

Charging Handle
(Correct terminology is Operating Rod, but one
could be excused for calling it the charging handle.)

To be fair, up to this point, every US Army service rifle had a reciprocating charging handle. Which means the handle is part of, or attached to, the bolt. When the bolt moves, so does the handle. Many guns are designed this way, like the M14 pictured here.

This feature allows the shooter to force the bolt into a closed position. It sounds good in theory, but shoving the bolt into place, if it doesn't want to go, isn't always the best idea, and it could cause more problems than it solves.

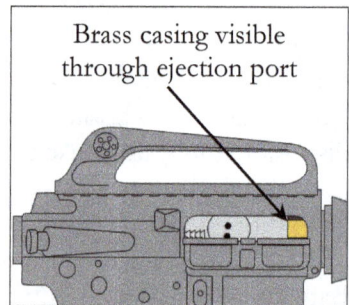

Brass casing visible
through ejection port

On the other hand, the reciprocating handle allows the shooter to perform a "press check" to see if the gun is loaded without ejecting a round.

To perform a press check, pull back slightly on the charging handle until you can see the brass casing, then push the bolt forward until fully seated.

Well, as you can imagine, Colt really wanted the contract to make guns for the US military and thus, the forward assist was born.

M16 A2
The only variant of the M16 that has a 3-round burst. All others can select to full auto

The forward assist is a button-like device on the right side of AR variant rifles.

The bolt carrier houses all the stuff that makes up a bolt, including the guts that make up the direct impingement system.

The forward assist is held away from the bolt carrier by a spring. This allows the bolt carrier to travel back and forth unimpeded.

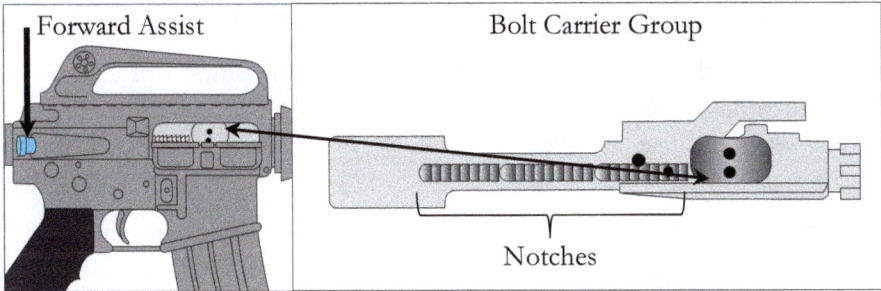

Forward Assist

Bolt Carrier Group

Notches

When the forward assist is pressed, it catches the notches on the side of the carrier, allowing the operator to positively seat the bolt.

The AR variants are the only rifles I'm aware of that use this feature. In fact, the original M16s didn't have the forward assist.

There is great and heated debate in the gun forums about whether a forward assist is needed or not. I stay out of all that silliness.

Yeah, TMI. However, if your character is using an AR variant, it may have a forward assist. It's just one of those things your character may need to interact with. Now you know why it's there and what it does.

Fun Fact: AR stands for ArmaLite Rifle, not assault rifle.

Carbines versus Rifles

The carbine was developed so cavalry could more easily employ firearms from horseback. In modern times, the carbine is the preferred size weapon for close-quarters battle (CQB) and building clearing.

A carbine is nothing more than a rifle with a shorter barrel. An M4 – with a barrel length of 14.5 inches, is a carbine version of the M16 – which has a 20-inch barrel.

There are both industry and legal definitions describing what a rifle, carbine, and short-barrel rifle are. The basic rule of thumb is:

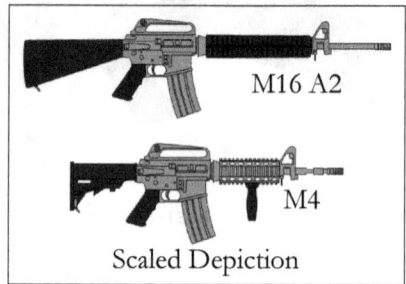

M16 A2

M4

Scaled Depiction

- **Rifle:** Barrel length of 20 inches
- **Carbine:** Barrel length between 16–20 inches
- **Short barrel rifle:** Barrel length less than 16 inches with an overall length less than 26 inches, or a handgun fitted with a buttstock.

Of course, there are exceptions to the above guidelines. For example, the Mini-14 is a rifle that can have a barrel length less than 20 inches. Rifle is used because it's *named* as a rifle. So, all carbines are rifles, but not all rifles are carbines.

My Advice: Go with what the manufacturer or common name is and try not to vary. If the gun is named as a carbine always use carbine. Don't use rifle, even though you know better.

Scenario: A soldier inspects her M4 carbine.

She picked up the M4. The new rifle was lighter and shorter than her old M16a2.

The above example is correct in all ways. However, some knucklehead out there is bound to try to snipe at the author.

"M4s are carbines, duh!"

Best to stick with always calling the weapon by its stated or manufacturer's name.

She picked up the M4. The new carbine was lighter and shorter than her old M16a2.

94

Bullpups

To be classified as a bullpup, the trigger and grip for the firing hand must be located in front of the weapon's breech. Firearms of this type have all the functional parts that regular rifles do; they're just in different locations.

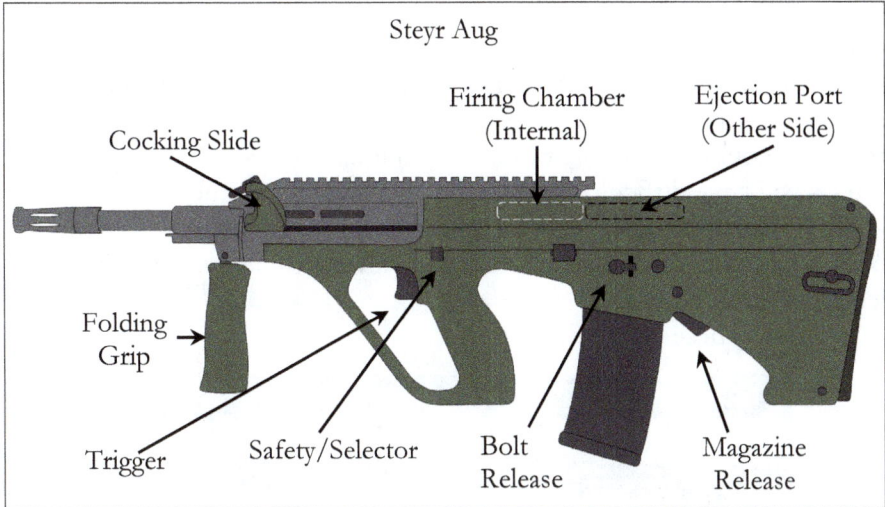

Steyr Aug

Firing Chamber (Internal) — Ejection Port (Other Side) — Cocking Slide — Folding Grip — Trigger — Safety/Selector — Bolt Release — Magazine Release

The simplified drawing below shows the basic operation of the bullpup. Green areas denote trigger operations. Blue denotes bolt operations.

The bolt and trigger assemblies are located in the stock, behind the actual trigger. They are connected to the operational parts by rods and linkages.

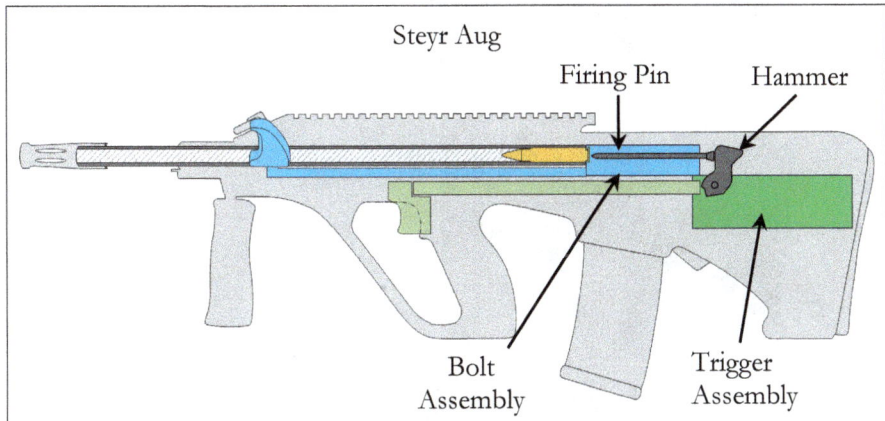

Steyr Aug

Firing Pin — Hammer — Bolt Assembly — Trigger Assembly

Video: https://www.youtube.com/watch?v=gIOUv7aFbto

The bullpup is a great idea, but I find actually operating this type of gun to be a bit awkward for a couple of reasons.

First, the ejection port is far back on the stock. This means that if you find the need to fire left-handed, hot brass will be ejecting into your face. Second, in this model, the magazine release is not in an easily accessible location to either the firing or the support hand. Ergonomically, I give one out of five stars.

However, with proper training and repetition, I'm sure the bullpup can be effective. (After all, the Steyr AUG is the primary service weapon for many countries and organizations.) But characters – even experienced ones – will have difficulty using bullpups for the first time.

The main advantage of the bullpup design is that it can reduce the overall length of the firearm to that of a carbine while retaining the barrel length of a full-sized rifle.

Consider these scaled depictions.

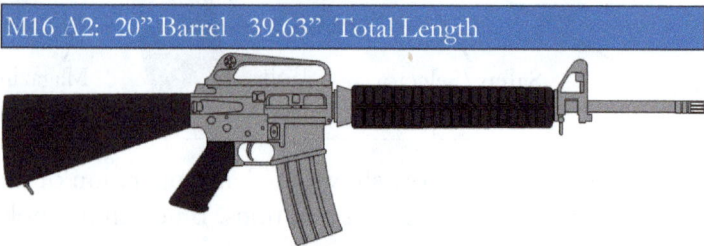

M16 A2: 20" Barrel   39.63" Total Length

M4: 14.5" Barrel   29.75" Total Length

Steyr Aug: 16" Barrel   28.15" Total Length

As stated earlier, the longer barrel length means better ballistics. The bullpup, with 1.5 more inches of barrel, has better range and a shorter overall length than the M4.

Machine Guns

Let's step back to the mid-1800s. Breech-loading firearms were taking their place in the hands of shooters across the globe, and reliable repeating rifles were on the horizon. What could be better than a repeating gun, a machine gun!

Richard Gatling invented a class of firearm that still bears his name. The original Gatling gun, 1861, was a hand-cranked, gravity-fed, multi-barreled gun.

Gattling Gun
© 2004 Matthew Trump

Each barrel had its own hammer and striker assembly and fired independently when rotated into position. This configuration gave the barrels a chance to cool while maintaining a high rate of fire.

The original Gatling gun doesn't really count as a true machine gun. For one thing, it's manually operated and energy from fired rounds isn't harnessed by the operating system. So, it can't be classified as an automatic firearm.

Because the Gatling gun is really a collection of guns constructed to fire independently, it's more of a weapons system.

However, its rate of fire and devastating battlefield effects would inspire the next generation of machine guns less than thirty years later.

Fast forward to 1959, and we can see the influence of Gatling's design in the M61 Vulcan and the M134 Minigun, which, apart from being operated electronically, are basically the same design.

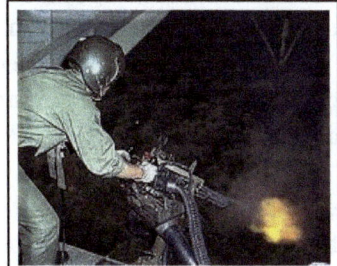

Minigun
Photo by: Paul Hagerty, SSgt, USAF

Fun Fact: The M134 gets its name *Minigun* from the fact that it's a scaled-down version of the 20 mm Vulcan cannon.

Maxim Gun, 1886

In 1883, Hiram Maxim was working with electricity when a friend told him,

> *Hang your electricity. If you want to make your fortune, invent something to help these fool Europeans kill each other more quickly!*[1]

The Maxim gun was the first true machine gun. It was a recoil operated, belt-fed, water-cooled machine gun. It's crazy to think that this machine gun predates functional automatic rifles.

Maxim Gun
Author: Zorro2212

The Maxim was great, but it was heavy. While it only required one man to fire, it needed a full team to help sustain it on the battlefield.

Armies needed a light machine gun that could be more easily repositioned on the battlefield … and mounted on planes.

Lewis Gun, 1911

The Lewis gun was a gas-operated, pan-fed, air-cooled machine gun, and it was used extensively by the British in WWI.

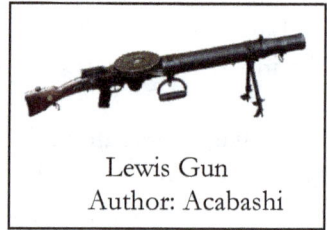

Lewis Gun
Author: Acabashi

Lighter than water-cooled guns, the Lewis could be mounted on aircraft and toted around on the battlefield. Perfect, right?

Not so much. For its day, the Lewis was a vast improvement over the water-cooled guns, but it couldn't sustain fire for too long before it overheated. The Lewis could only fire twelve 47-round magazines in a row before it needed a break.

Clearly, firing sustained bursts through the gun overheated the barrel. Closed bolts don't allow for the circulation of air through the barrel and chamber. Thus, heat builds up rapidly but dissipates slowly.

A new design was needed.

---

[1] John H. Lienhard, "Hiram Maxim," in *The Engines of Our Ingenuity*, no. 694, Cullen College of Engineering, University of Houston, https://engines.egr.uh.edu/episode/694.

Interlude: Firing from the Open Bolt

Open bolt design is generally found in firearms designed to fire full-auto – machine guns and submachine guns. These guns stay cooler longer because the open bolt allows air to circulate through the barrel and chamber when they aren't firing. This simplified, generic graphic explains how this design works.

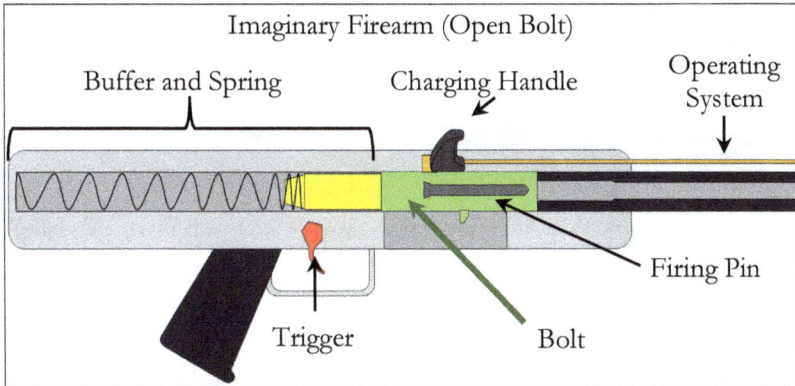

Imaginary Firearm (Open Bolt)

- The bolt is locked to the rear, setting the trigger and compressing the buffer spring. Ammunition is loaded.
- When the trigger is pulled, the bolt moves forward, picking up the first round.
- The round is fired as soon as the bolt locks into place with the chamber.
- The operating system forces the bolt back, extracting the empty casing.
- The empty casing is ejected. The gun is ready to fire again. The weapon will fire as long as the trigger is depressed.
- Firing stops when the trigger is released.
- Open bolt guns go thunk after the final round, unless the firearm is equipped with a bolt hold open device.

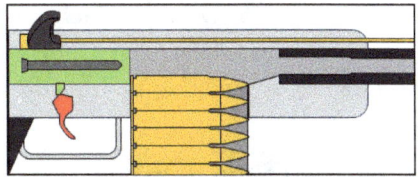

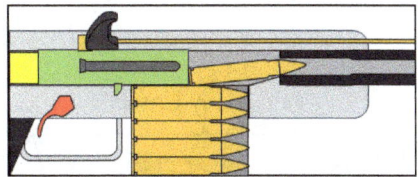

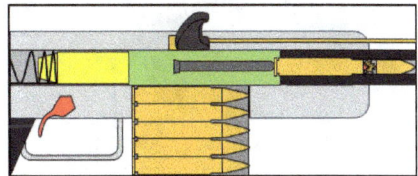

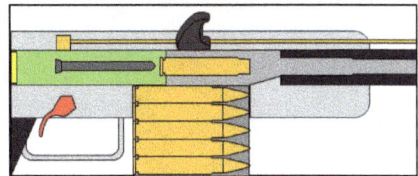

99

MG 34, 1936

Even if you have a machine gun that fires by means of the open bolt, you can't fire the weapon indiscriminately. Burning through hundreds of rounds in one continuous burst is bad for the gun. As the barrel heats up from sustained use, it suffers wear at an accelerated rate. In extreme cases, the barrel can warp and melt.

So, we need a gun that fires from the open bolt, and we need to be able to change barrels quickly ...

Author: Armémuseum (The Swedish Army Museum)

The MG 34 and its younger brother, the MG 42, met those criteria, and they ruled the battlefield in WWII.

The MG 34 was a gas-operated, belt-fed, air-cooled machine gun capable of firing more than 800 rounds per minute.

No other army had a machine gun that was as light, durable, and reliable as the German machine guns. In fact, the MG 42 is still in service to this day and elements of its design have been incorporated in many other guns.

Anatomy of the Belt-Fed Machine Gun (M60)

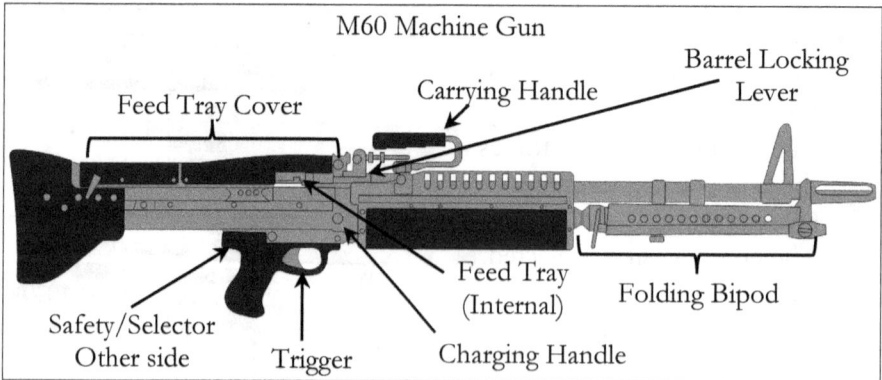

M60 Machine Gun

Feed Tray Cover

Carrying Handle

Barrel Locking Lever

Feed Tray (Internal)

Folding Bipod

Safety/Selector Other side

Trigger

Charging Handle

So, you want to add a machine gun to your character's arsenal? You'll want a light machine gun that can be easily managed by one person.

Not to say that a light machine gun is light. The venerable M60 weighs over twenty-three pounds. You're also going to need ammo – a lot of ammo – which also contributes to the load your character must carry to be effective. All that aside, let's look at what you need to know.

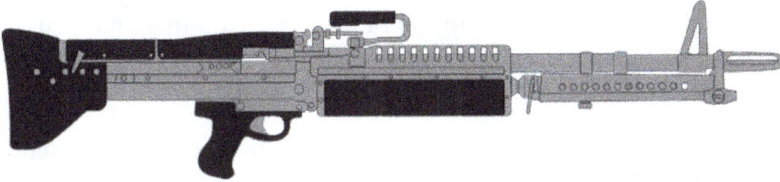

Most machine guns are belt-fed. Disintegrating ammunition belts are made up of small links that fit around the cartridge and link together. As the ammunition is fired, the links and expended brass pile up around the gun.

Non-disintegrating belts can be made of canvas, plastic, or flexible metal. These were a common source of feed for machine guns in the early to mid-1900s, and they could be easily reloaded by gun crews. The problem with a belt like this is that the used portion can't be discarded when the gunner needs to move. This represents a tripping hazard for the gun crew. For this reason, most modern machine guns use disintegrating belts.

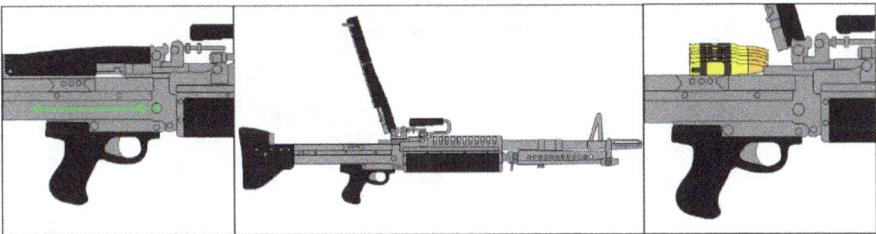

To Load:
This procedure is typical for most machine guns.
- Using the charging handle, retract the bolt until it locks in place. Return the charging handle to its original position and ensure the weapon is on safe.
- Open the feed tray cover.
- Place the belted ammunition on the feed tray and hold it in place as you close the feed tray cover.

The rate at which the machine gun is fired depends on the gun, ammunition, situation, and doctrine of the military in question. Rates of fire are used as a guideline for barrel temperature management.

There are three rates of fire for machine guns:

Cyclic fire is the mechanical rate at which the gun operates. This doesn't take into account ammunition limitations or stress on the gun due to heat buildup. When you look up the rate of fire for a machine gun, this is the number you see. The M60 has a rate of fire between 500–650 rounds per minute.

Sustained fire is the rate at which the gun can be fired for an extended period of time. This takes into account heat management and barrel care. Sustained fire is unique to each gun. However, most military doctrine recognizes the sustained rate of fire at 100 rounds per minute.

Notice the rate the gun is capable of versus the rate at which it should be fired. To regulate fire, gunners are taught to fire in bursts, usually three to five, or six to nine round bursts. This technique helps keep the rate of fire at the sustained level.

Rapid fire is a rate of fire that can't be maintained for long without doing harm to the barrel. Rapid or suppressive fire is used against imminent targets that deserve extra attention, such as a moving vehicle, aircraft, or a group of combatants.

Regardless of the rate of fire, at some point the barrel will need to be changed to preserve the gun. Your characters need to take care when operating around hot barrels. I grabbed a hot barrel in the dark once. Fortunately, it was heavily oiled, and my hand didn't stick to it. Otherwise, I would have had third-degree burns. I give that experience zero stars, would not recommend.

Note: Back when the M60 was still in service, gunners would use an asbestos glove to handle the hot barrel during a change. Fortunately, both the M249 SAW and the M240b have carrying handles attached to the barrel, which removes the need for a cancer glove.

When in a static or defensive position, machine guns are set up to provide as much support and coverage as possible.

The typical US Army infantry patrol base is set up in a triangular formation with the squads forming the sides and the machine guns at the corners.

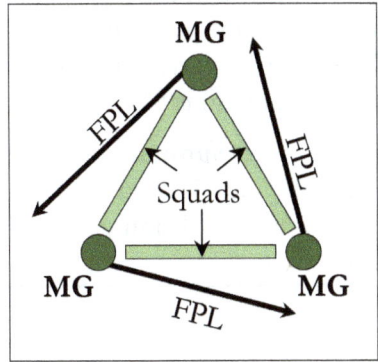

Each position has sectors of fire that interlock creating a strong defense. Interlocking sectors are also used in trenches and forward operating bases (FOBs).

If a position or section of the defense is in danger of being overrun, gunners are told to fire on the final protective line, or FPL. At that moment, every gun it that section goes cyclic.

Army doctrine calls this *final protective fire*. Rangers refer to this as *death blossom*. All guns are fired without regard for ammunition consumption or damage to the gun.

Talking the Guns:

Talking the guns refers to a technique used when a unit has more than one machine gun. The gunners take turns firing. This keeps up a sustained rate of fire on a target and allows the guns to fire longer without needing a barrel change.

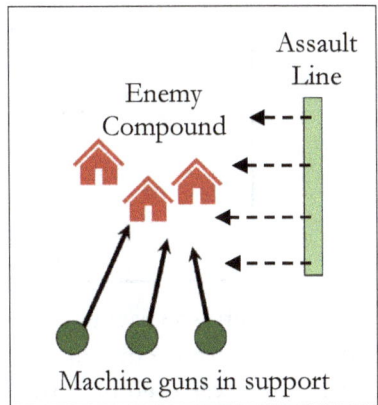

It's like a conversation between guns. The first gun speaks, the next one in line answers. And so on, until it's time for the first gun to speak again. If the gunners are good, the rate of fire they can put on a target is staggering.

This technique can be used both offensively and defensively.

Video: https://www.youtube.com/watch?v=ThsGaj6_Cpg

Submachine Guns

A submachine gun is usually a shoulder-fired rifle, carbine, or short-barreled rifle designed to fire handgun ammunition. The acronym for this type of firearm is SMG, which I'll use for the duration of this section.

Thompson Submachine Gun, 1921

The infamous Thompson, or tommy gun, was the first fully automatic gun to use pistol ammunition. It fires .45 ACP ammunition from the open bolt.

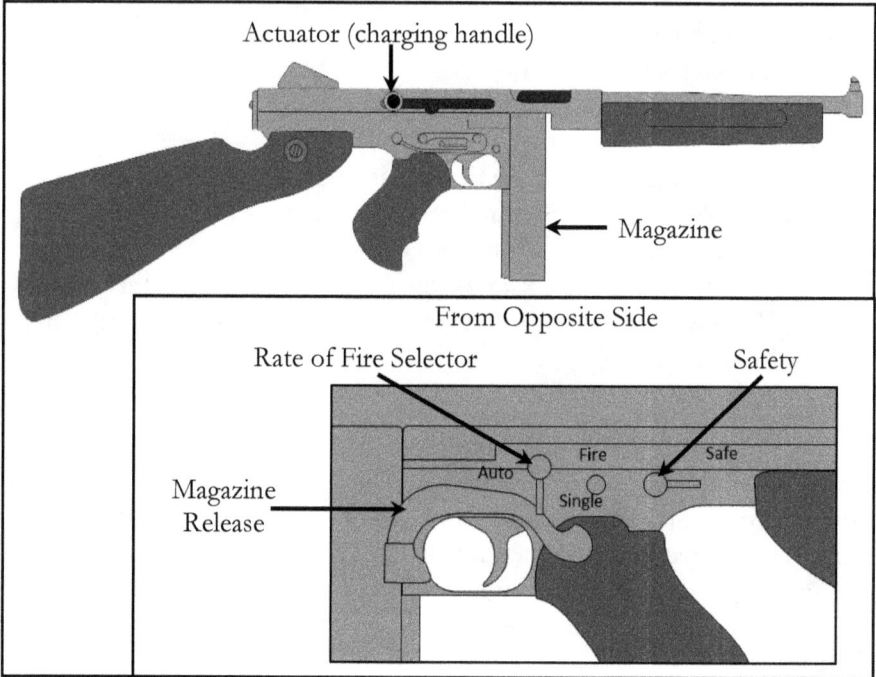

The Thompson saw many variations over the years, and some models had the actuator on the top of the gun. The Thompson was extremely popular and was used by dozens of countries all over the world. The model pictured here is the M1A1 used by the US in WWII.

The M1A1 has an unusual two selector system. One lever controls the rate of fire, while the other engages the safety.

It was fed by twenty or thirty round box magazines, or by one hundred round drum magazines.

Since the Thompson has several possible variants, be sure to distinguish which model your character is using if you choose to use this firearm.

SMGs are also known as machine carbines, just to add confusion.

Uzi, 1954

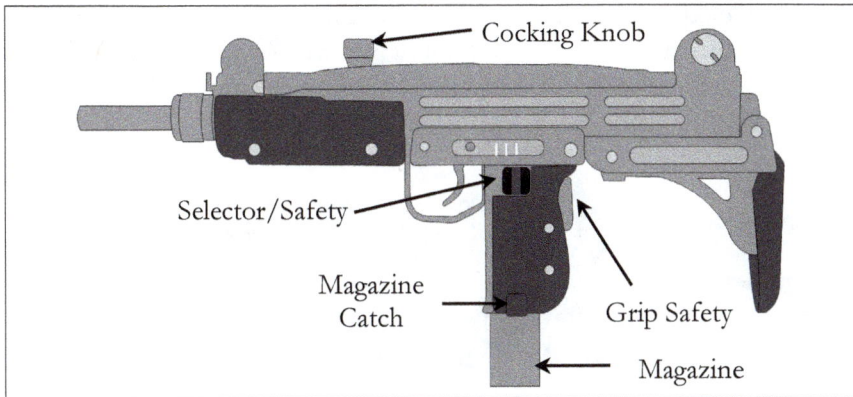

The Uzi is one of the most iconic of the SMGs and is indicative of machine carbines that fire from the open bolt. (Page 99)

New, civilian models are called the Uzi carbine. They have a longer barrel to meet barrel length requirements of US law and they fire from the closed bolt. However, sale of the Uzi has been prohibited in the US since 1989.

Since the Uzi has several possible variants, be sure to distinguish which model your character is using if you choose to use this firearm.

MP5, 1966

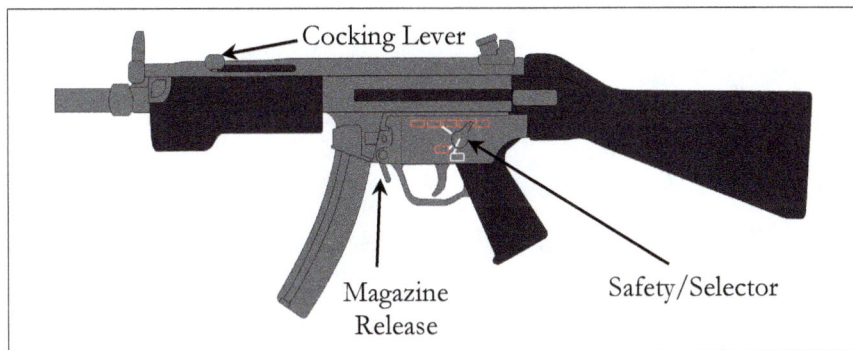

The MP5 is another one of those iconic guns that had a global impact. It was the preferred firearm of almost every Special Operations unit until the advent of the M4 carbine. It's designated as a submachine gun and has no carbine or pistol appellation.

Here's where the SMG designation gets really confusing. The SMG can be modeled on any platform. You like the way the AR platform works? Slap a 9mm barrel on it and now you've got an AR SMG. Of course, you'll need to get new magazines and make some other adjustments, but that's basically it.

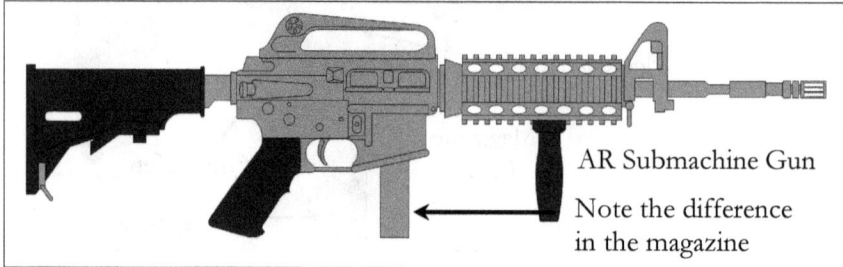

AR Submachine Gun

Note the difference in the magazine

Merriam-Webster defines SMG as: a portable automatic firearm that uses pistol-type ammunition and is fired from the shoulder or hip.

Note the definition doesn't specify full- or semi-automatic fire.

Advice: Only use the SMG designation on weapons called submachine guns. Stay away from the oddballs and keep your selection common and recognizable.

You could write in an AR 9mm SMG – they do exist – but it might serve you better to go with an UZI or MP5. Just remember that some of these guns fire from the open bolt.

Interlude: Pistol-Caliber Rifles and Rifle-Caliber Pistols

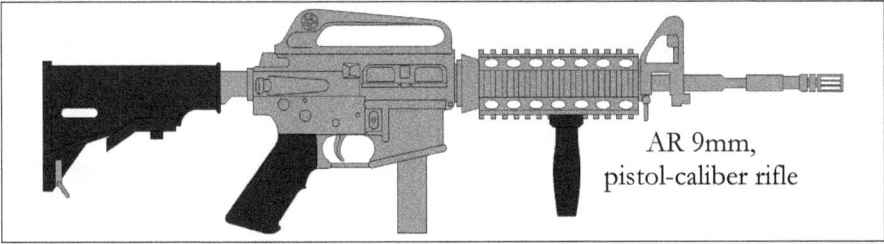

AR 9mm,
pistol-caliber rifle

I mentioned above that a rifle that fires handgun ammunition can be called a submachine gun. It may also be called a pistol-caliber rifle. And on the previous page, Merriam-Webster defined SMG as: a portable automatic firearm that uses pistol-type ammunition and is fired from the shoulder or hip. The following rules will help with this conundrum:

- If the manufacturer calls the firearm a submachine gun – call it a submachine gun.
- If the rifle is chambered in a handgun caliber and can fire in a fully automatic mode – submachine gun.
- If the firearm is chambered in a handgun caliber and called a rifle – call it a rifle.

Rifle-caliber pistols are a problem too. These are rifles with a barrel length reduced to below carbine lengths. These guns don't have a stock, although, they may be fitted with a "pistol brace."

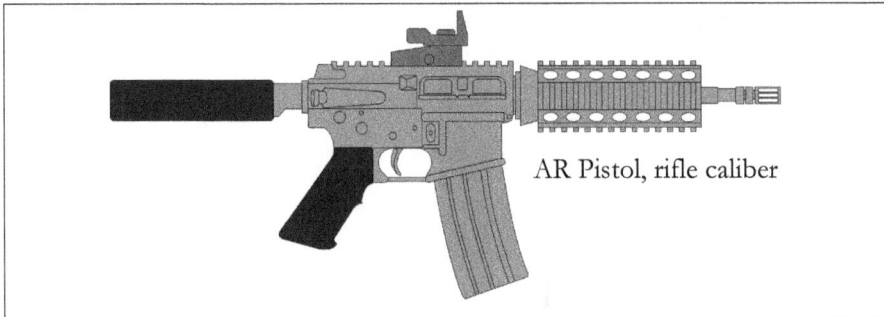

AR Pistol, rifle caliber

This is an attempt by gun enthusiasts to sidestep the Bureau of Alcohol, Tobacco, Firearms, and Explosives (BATF). By calling the firearm a rifle-caliber pistol, they avoid having to register the firearm as a short-barreled rifle and avoid paying the extra costs to legally own it.

With the plethora of firearms available, I suggest avoiding these guns unless they are important to plot or character. Instead, give the character a short-barreled rifle and move on.

Short-barreled rifles (SBR) lose a lot of range and velocity due to the shorter barrel length, but these guns are designed to be used in CQB or close-quarters combat (CQC). Maneuverability is the primary advantage of the SBR.

Moving through a floorplan with a long-barreled weapon can be difficult, which is why MP5s were so popular with Special Operations units back in the 1980s.

MP 5

Reducing the length of the barrel also allows the operator to install a suppressor, bringing the gun back to a carbine length. This is another huge advantage during operations where stealth is required. Anyone who's whacked tangos playing *Rainbow Six* can attest to this.

M4 Suppressed
with Optic

To avoid the bother of explaining an AR pistol, I'd describe the firearm:

The villain removed the stock from his short-barreled AR and put on a suppressor. Slung up into his armpit, it was perfectly concealed by his coat.

Timeline of Rifle Development

Sharps Carbine – 1848: Still used paper cartridges

Centerfire Cartridge – 1852: The first modern cartridge

First Falling-Block Rifle – 1862: Peabody-Martini

First Lever-Action Rifle – 1862: Henry rifle

First Rolling-Block Rifle – 1865: Remington

First Bolt-Action Rifle – 1866: Frederick Vetteri

Conversion of Muskets – 1873: US – Springfield Trapdoor

First Box Magazine – 1879: James P. Lee

First Cartridge Clip – 1885: Ferdinand Ritter von Mannlicher

First Automatic Rifle Patents – 1885: Hiram Maxim

Automatic Rifles Become Common – 1920s/1930.

The early 1900s saw an explosion in design ideas for the automatic rifle. There are too many to list here. There are also naming conventions that affect "historical first" internet searches. Use the following guidelines when referring to automatic rifle types.

Automatic Rifle: Select fire rifle that fires rifle ammunition.

Submachine Gun: Select fire, shoulder fired rifle, or carbine that fires handgun ammunition.

Machine Gun: A weapon designed for sustained automatic fire of rifle ammunition.

A quick note on the AR series rifles. This includes AR-15/M16 variants. AR stands for ArmaLite rifle, not assault rifle.

ArmaLite sold Eugene Stoner's design to Colt, who produced the M16 and turned the weapon, for good or ill, into the icon it is today.

Any weapon that uses the same design basics as the M16 is classified as an AR.

Note: *Assault rifle* is a political term. Back in WW1, when bolt-action rifles were issued to the military, bolt-action rifles were assault rifles. The same goes for any other firearm. So, unless it's important to character development or the story, stay away from this politically charged term.

Operating Long Guns

How the Longarm Is Carried

Okay, time for some fun. Oh, and no more muskets. Everything from here on is going to be about modern rifles and shotguns. Let's start with how the weapon is carried. Modern technique calls for the weapon to have a fully loaded magazine, a chambered round, and be on safe. Or locked and loaded.

"Lock and load." I love that phrase. When you get that command, you know it's time to dance. It comes from, damn it, the flintlock loading procedure. Go to half cock – lock, and load. Anyway, your basic cop, soldier, or bad guy is going to be carrying the rifle like this: loaded and on safe.

Scenario: Modern-day S.W.A.T. team is about to perform an entry on a possible terrorist hideout.

"Thirty seconds," the team leader called. "Lock and load, weapons hot!"

Unfortunately, we have another error, sort of. *Weapons hot* means, go off safe. There was a time when operators performed raids and entry actions with their weapons on fire. This causes most modern professionals to shudder.

An entry or raid is controlled chaos. At any point during the action, an operator could go from firing their weapon to grappling with a suspect or giving aid to a victim. The last thing anyone wants in the last two activities is an accidental or negligent discharge.

Modern operators keep their weapons on safe until they're ready to fire. Once the shooting is finished, they immediately return the weapon to safe. So where did this practice come from? Poor engineering, that's where.

Back in the 1980s, the MP5 was the preferred weapon of anti-terrorist and S.W.A.T. teams because it's light and small. However, the MP5 did have one flaw. It was ergonomically unfriendly for safety manipulation.

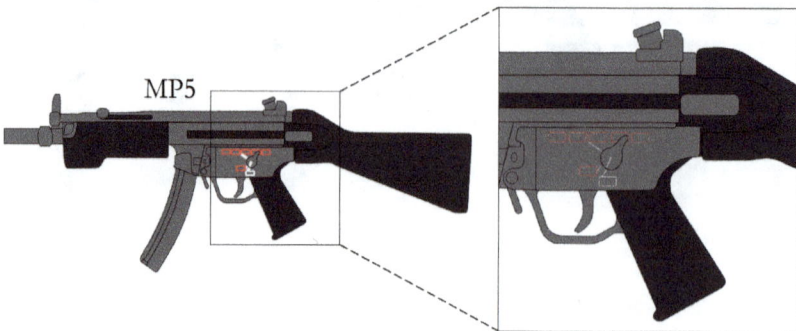

MP5

The operator had to alter their grip with the firing hand to be able to reach the selector. This put the trigger finger in an awkward position to engage rapidly. The selector has since been redesigned and now allows for rapid manipulation.

Given that modern training techniques involve safety manipulation, the average "informed" reader will question the "going hot" order, regardless of the era or weapon.

The "lock and load" command is good. This is the command to chamber a round and make the weapon ready to fire. Some may question why the author waited until the thirty-second mark. Policy or safety rules for an organization may dictate when to chamber a round.

Firearm Storage versus Staging

Stored versus staged is one of those semantic issues. The argument goes that if the gun is stored, it's unloaded and boxed away or in a safe.

If it's staged, the gun is readily available if needed, possibly in a gun safe, locking mechanism, or freely available. A staged firearm can be in any condition from fully loaded to completely unloaded with ammunition readily available.

I guess I can see the argument. There is a difference. However, I'd write my way around the issue.

She kept a loaded pistol in her nightstand.

The empty shotgun sat in the closet.

He kept a loaded pistol in the safe with his money, just in case …

I'm not going to go into a huge dissertation about how a person should keep their guns in the real world or where they should be staged. All I'll say is that consideration should be given to whether there are children around and if a guest can access the firearm. Given those variables, the weapon can be in any of the following loaded conditions:

- Fully unloaded.
- Loaded and on safe.
- Loaded, on safe, and trigger locked. A trigger lock is either a feature of the gun (pistols) or an external lock that fits around the trigger guard and prevents access. Very handy if there are kids in the house.
- Cruiser-ready. This is a term that means the magazine is full – and inserted if the firearm loads from an ejecting magazine – but the chamber is empty. All you need to do is grab the gun, chamber a round, and get into the action.

Cruiser-ready is a term that comes from the way police store rifles and shotguns in their vehicles. Police cruiser – cruiser-ready. This is a great way to store a gun you may need to use in a hurry. As an added detail for shotguns, the action must be free to operate. This means either the trigger has been pulled, or the slide-action lever has been activated.

This is a great condition for home defense. Nothing says, *GET OUT,* like racking a shotgun.

Carrying Multiple Firearms

Primary firearms are usually longarms – rifles or shotguns that either do the most damage or have the greatest capacity. This will be the type of gun in your character's hands when they go into harm's way.

Secondary firearms are backup guns for use when the primary weapon fails. This is usually a pistol. If the character is only armed with a pistol, that is their primary firearm.

Tertiary firearms are used for special tasks. They're usually grenade launchers or shotguns and are used to launch tear gas or breach doors. These are carried in a loaded condition to meet the task.
- Grenade launchers will be loaded and on safe.
- Breaching guns will be cruiser-ready. (Page 113)

Soldiers and police officers that use shotguns to breach doors often carry them in cruiser-ready condition. Shotgun breaching is cool and fun. As the breacher on my old team, I got to perform this role many times. It always put me where the action was.

Reload versus Transition

When armed with both rifle and pistol and the rifle runs empty, the character will either perform a reload or transition to the secondary firearm based on the situation. This is a subjective call, but as a rule of thumb, twenty-five yards is the generally accepted distance.

Emergency reloads happen when there is time and distance – the opponent is outside twenty-five yards, or the character has time and cover. Cover is an object between combatants that can stop bullets. The character will:
- Eject the empty magazine.
- Insert a fresh one.
- Manipulate the bolt to chamber the first round.

Note: The conditions above also apply to the decision to reduce a malfunction (clear a jam, Page 220)

Transitions – from rifle to pistol – occur when the character has very little time and the opponent is close. Steps to perform the transition:
- The shooter attempts to put the rifle on safe.
- The rifle is slung out of the way while the operator simultaneously draws the pistol.
- The rifle will be put back into operation as soon as conditions permit.

Note: This also applies to malfunctions and stoppages. (Page 220)

Interlude: Shotgun Breach

There comes a time in everyone's life when you want to get through a door, but you just don't have the key.

To breach a door using a shotgun, the first thing to know is that you shouldn't use slugs or buckshot. Those types of ammunition do a lot of damage and could injure noncombatants inside the room. Use bird shot or special breaching rounds.

The target of the breach isn't the knob; it's the latch mechanism. Place the barrel of the shotgun between the knob and the edge of the door. It should be slightly above the centerline of the doorknob and angled downward and toward the door frame at a 45° angle. This will take out the latch and the part of the door frame that houses it.

- Zatoichi1564

Procedure:

- The breacher is called to the locked door.
- The breacher slings their primary weapon and retrieves the shotgun.
- The breacher moves to the door, racks in a shell, targets the latch, and looks at the number one operator.
- The number one operator gives the nod and the breacher fires, kicks the door, and moves to the side.
- The breacher stows the shotgun, retrieves their primary weapon, and enters the room or maintains security.

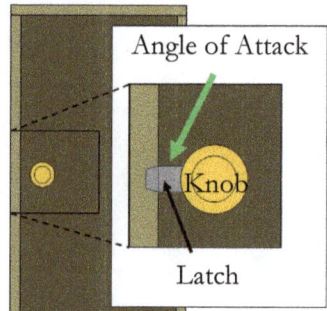

Angle of Attack

Knob

Latch

There is no need to safe the shotgun because there's an expended shell in the chamber. If the breacher gets excited and chambers another round, the weapon must be put on safe.

- Justin Young

Sighting Mechanics

Before we get into the sighting systems available for rifles, we should take a quick detour into ballistics.

And before I do that, I'm going to warn you that this gets uber nerdy.

Rifles are very different than handguns in that they aren't ready to use right out of the box. They must be sighted in on a range first. The procedure for doing this is called zeroing, and it's the same for every sighting system.

All shooters that use a rifle will zero that rifle at some point *before* they go off to do their thing. Zeroing is adjusting the sights so that the point of aim and point of impact are the same at a specific range. Your rifle toting character will be familiar with this process.

You will probably never write about the procedure, but it can impact your character's ability to hit what they're shooting at if they use a rifle other than their own.

Scenario: The hero's rifle is damaged in an explosion. He finds another and gets back in the fight. Both rifles use mechanical sights.

Sergeant: Damnit, Soldier, why can't you shoot that guy?

Hero: I'm trying to, but this ain't my gun.

Sergeant: Where the hell is yours?

Hero: Inoperative. I'm using Smitty's.

Sergeant: Well, shit, do your best.

The hero can't hit the bad guys because he's using a rifle that's been sighted and zeroed for another person. You can take my word for it and skip ahead, or you can read on and discover why the hero can't hit his target.

Bore/Sight Relationship

These graphics are exaggerated for clarity.

I'll start with iron sights, which are simple metal sights attached to the gun. Generally, front sights adjust elevation, up and down, and rear sights adjust windage, left and right. However, this isn't always the case.

Bullets, when fired, follow a parabolic trajectory caused by the relationship between the sights and the center of the barrel or bore. The distance between the two is called the standoff. The greater the standoff, the more drastic the arc.

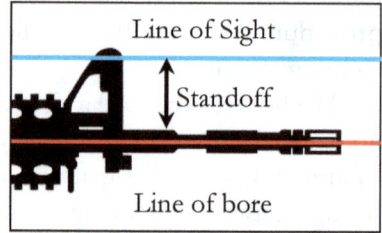

This holds true for every sighting system, be they iron sights, scopes, lasers, or any other kind of optic.

So, we have a point of aim – where we want the bullet to go, and a point of impact – where the bullet actually goes.

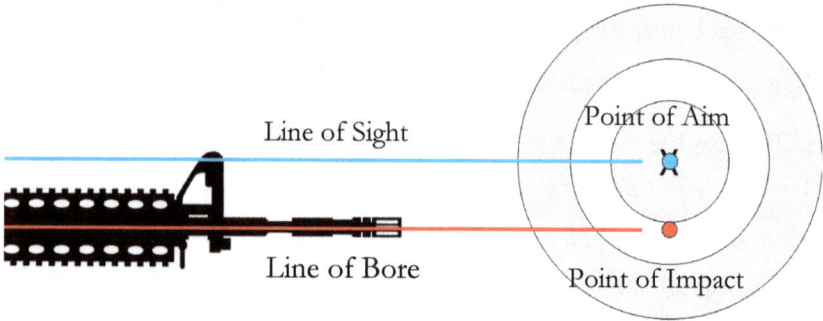

If the gun is held level, and there have been no adjustments to the sights, the strike of the round will be lower than the line of sight by the same measurement as the standoff.

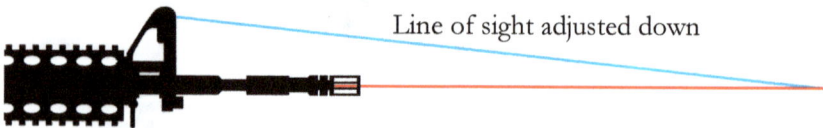

To fix this problem, we must adjust the sights so the point of aim is the same as the point of impact at the desired distance. Adjusting the front sight down raises the point of impact when aiming at the same place on the target.

Because the barrel is now canted slightly upward, the trajectory of the projectile becomes parabolic. This is where the term *bullet rise* comes from.

Same point of aim.
Point of impact is raised

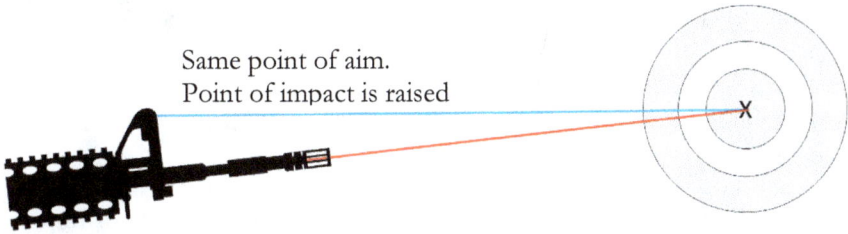

The distance at which the point of aim and point of impact are the same is called a zero. If I achieve a zero at twenty-five meters, I have a twenty-five-meter zero. Of course, there's another zero at a further distance. This is because gravity is a cruel mistress, and the bullet crosses the line of sight again on its way back to the ground.

So, in a perfect environment, we can predict where the bullet will be relative to the line of sight anywhere along its trajectory.

• Inside and outside the zero distances, impact will be low.
• Between the zeros, impact will be high.

Apex

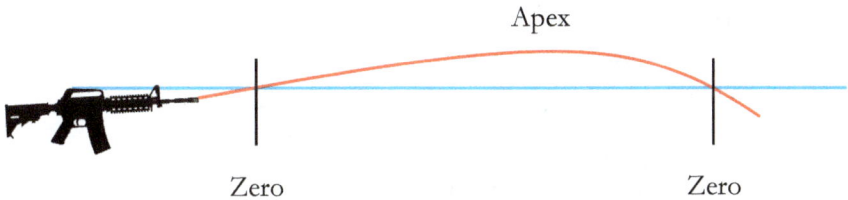

Zero   Zero

Sorry, we're not done yet.

Because people have different physical features and their faces rest in different positions on the stock, their eyes line up differently with the sights. Don't even get me started on astigmatisms and short-/near-sightedness. Suffice to say, there will probably need to be left and right adjustments to the iron sights as well.

Thus, even if people carry the same firearm and acquire a zero at the same distance, the adjustments made to the sights could be radically different between them.

This is why our hero couldn't hit a damn thing in our last scenario.

The Zeroing Procedure

To zero, shooters must be able to have consistent grouping. If they can't keep their shots together, they can't get an average point of impact. This means they can't zero with any reliability. So, if the grouping is good, they continue with the procedure.

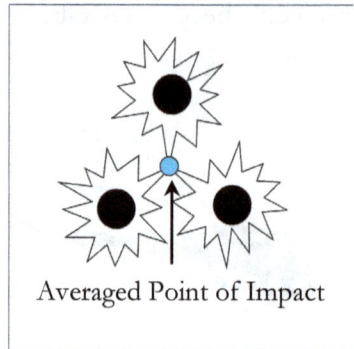

Averaged Point of Impact

The phrase "Tighten up your shot group," also means "Get your shit together."

Before the shooter can begin the procedure, they need to make sure the sights or optics are set at their mechanical zero – flush and center for iron sights and factory setting for scopes. Once that's done, they can start the process.

- The shooter fires three rounds, then checks the target.
- The shooter notes the distance from the shot group to the center of the target, then adjusts their sights accordingly.
- They fire three rounds to see the effect of their adjustments.
- They repeat this process until the point of aim is the same as the point of impact.

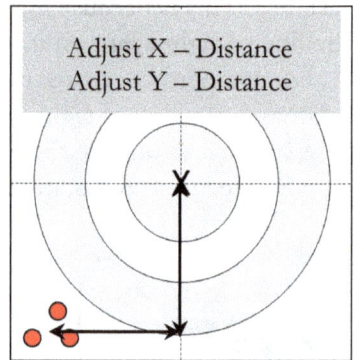

Adjust X – Distance
Adjust Y – Distance

The sum of all adjustments is called D.O.P.E, data on personal equipment. Snipers also use the term but with a different definition – data on previous engagement. They keep this information in a logbook.

I truly hope you never write the zeroing process into one of your books, but if you do, you'll need to be familiar with the proper terminology. Hell, you'll need to know some of these terms anyway if you're looking into scopes and optics.

Here's where it gets even more nerdy, a.k.a. deeper explanations than what's in the glossary.

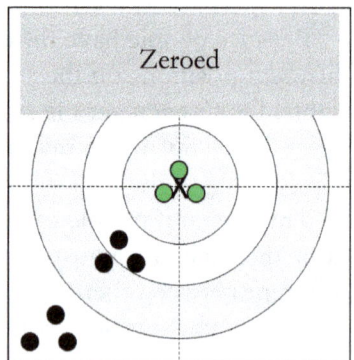

Zeroed

The sight radius is the distance from the back of the rear sight to the back of the front sight. Firearms with a longer sight radius are more accurate at further distances because a minor shift in orientation has less of an impact over distance.

It's called a radius because, when you move the barrel left or right, you're drawing an invisible arc that starts at the back of the gun.

This arc is affected when the sights are adjusted. If I move the rear sight to the right, the bullet impact will shift to the right when I line up the sights on the target.

We could measure the angle in degrees, but that unit of measure is too big for our purposes. Instead, we use minute of arc (MOA).

- 60 MOA = 1 degree.

Minute of arc adjustments get bigger with distance.

- 1 MOA = Adjusts 1 inch at 100 yards.
- 1 MOA = Adjusts 2 inches at 200 yards.

So, the further out you zero your sights, scope, or optic, the more accurate your gun will be at all distances.

At this point, I'm going to mix in optics. Optics differ from iron sights in that the sight picture through the device doesn't rely on face position. If the shooter can see through the device, and the device is zeroed, they can reliably use the firearm.

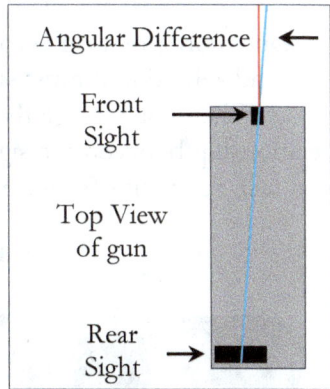

Angular Difference

Front Sight

Top View of gun

Rear Sight

Optical aiming devices use a fraction of MOA per click to adjust the aiming point, which corresponds to a measurement on the target at distance. Let's go back to our 100-yard target for a moment.

If my scope adjusts at ¼ MOA per click, and my shot group is 4 inches low and 3 inches to the left of the bullseye, then I need 12 clicks right/16 clicks up to be zeroed.

Different optics use different click values, and they vary between devices.

TMI, yeah. But you're still reading so you must want to know this stuff.

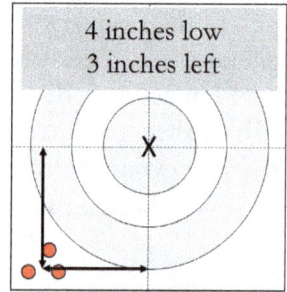

4 inches low
3 inches left

The other way optics use MOA is to describe how much of the target is covered by the sight's aiming point.

If your red dot optic is 3 MOA, it will cover a 3-inch circle on your target at 100 yards, 6 inches at 200 yards.

Red Dot Optic

One more point on zeroing. If you've mounted a scope or optic on a gun and zeroed it, it must stay in that position for the zero to remain good. When you change the position of the sight, you change the relationship between the sight and the bore. This changes the zero. You can take the sight off, but make sure you put it back in the same place.

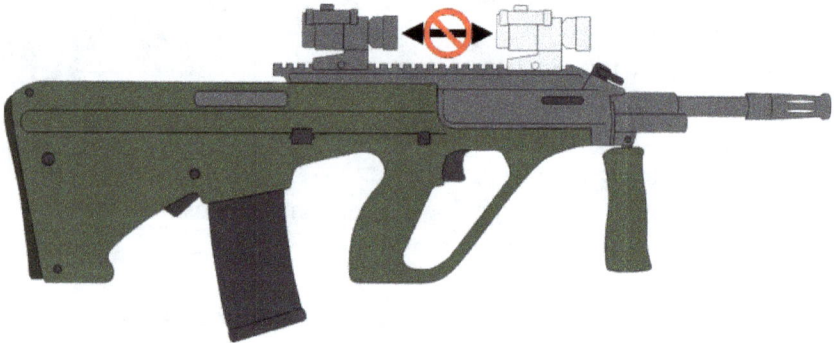

Sights

There are five categories of sites for firearms, and each category has its own variations. There is maximum opportunity to nerd out on how your character's gun is equipped. However, my standard advice applies:

If a firearm detail isn't critical to character, story, or plot don't even mention it.

Fixed Sights

Fixed sights are parts of the gun used for aiming. They're installed when the gun is manufactured and cannot be altered or adjusted. These sights are found in pistols and shotguns.

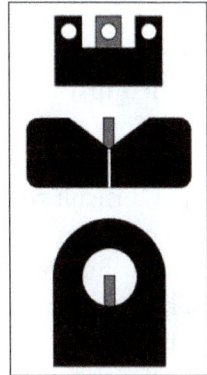

The bead sight on a shotgun has no zero distance. Where the slug or shot pattern impacts is completely dependent on how the shooter uses the firearm.

Fixed pistol sights are generally factory-zeroed for seven yards. They come in a multitude of varieties, including sights that glow in the dark.

Adjustable Iron Sights

Adjustable sights come in many flavors from blade sights, like on pistols, to peep sights. As the name implies, these sights can be adjusted so the shooter can zero the firearm.

Red Dot Sights

Red dot sights are a form of optic that can be powered by batteries, ambient light, or tritium. Often called holographic, or holo sights, these devices have become popular in the shooting community. They can be mounted on any firearm – including pistols – with the proper hardware. These sights must be zeroed when installed.

Red Dot Optic

Laser Sights

Nothing says, "You're in trouble now," like a little red laser centered on your chest. Laser sights come in a variety of colors and infrared and can be installed on both handguns and rifles. Laser sights must be zeroed.

Remember, lasers point both ways. That means your opponent knows where you are!

Scopes

Rifle scopes come with varying degrees of magnification and a ton of options in reticle design.

Some scopes are equipped with multiple hash marks or dots. These can be used to estimate size and range to a target. This takes practice and a little math. Your character is not going to excel at this skill on their first try.

Scopes are best used for distant targets. Magnification and field of view issues make them difficult to use against moving targets or targets at close range.

You can name the manufacturer of the scope, Leupold, Barska, Zeiss, but the shooting community is full of biases and there are preferences. Talk to an expert about what they prefer.

In addition, now that you've named the optic, if you talk about the reticle, you better get those details right. Magnification matters too. Certain magnifications are better than others at specific ranges, and the higher the magnification, the narrower the field of view. One last point: if you've named the organization a character works for, make sure to give them the proper equipment.

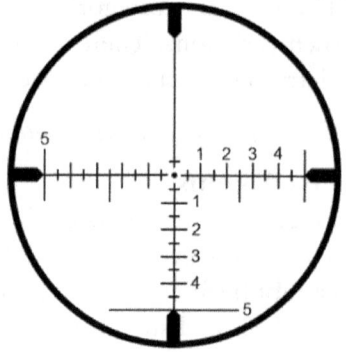

Backup Iron Sights (BUIS)

So, you took off those clunky iron sights and put on a super cool optic, but you gotta think about contingencies. Batteries fail and scopes are slow to engage within certain distances. For this, we have the BUIS. These sights are mounted on rifles that use optics as their primary sighting tool, and they can be of fixed or flip-up design.

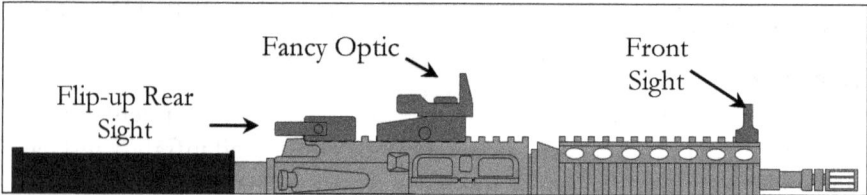

General Optics

Unlike iron sights, optics are good for any shooter once they're zeroed to the gun. The optic is in a fixed position on the gun, and the shooter needs only to be able to see the dot or reticle.

For those who skipped ahead, people have different physical features and their faces rest in different positions on the stock. This means their eyes line up differently with the sights. Iron sights must be zeroed to the shooter to ensure accuracy.

Bottom line: in an emergency, if your character picks up another character's rifle/carbine, and it's equipped with an optic, they're good to go. If the weapon only has iron sights, they may not hit shit.

Rifle Grab Bag

Welcome to the land of misfit toys. This is all the rifle/weapon stuff that doesn't really fit into any of the above categories. Some of these tips and tricks apply to pistols as well.

Sight Standoff

Remember the difference between the level of the sights compared to the level of the bore? The standoff can affect your character's ability to hit their target. Especially when shooting from behind cover. There are times when you can see your target through the sights, but you shoot the thing you're hiding behind. This can be dangerous and/or embarrassing.

Story Time

I was out at the range with a police agency, and we were doing cover drills behind vehicles. One of the guys forgot about the standoff of his sights and blew out the windshield of the police cruiser. I found the incident amusing and a great teaching point. The chief of police was not as enthusiastic.

Loading Magazines

You'll often see characters, especially in war movies, tap their magazines against their helmet before inserting them into their guns. It's not for luck. This action ensures that the cartridges are flush up against the back of the magazine.

If the cartridge is slightly forward, there's a small possibility that it may not feed correctly. Tapping the mag, if you have time, is just one more way to ensure your firearm will work.

Loading/Plussing Up

When your character loads their rifle or handgun, they can plus up one more round. You may see this in a firearm's capacity information.

Capacity: 15 +1

To achieve the plus one, load the gun as you normally would. Then eject the magazine and top it off with one more round.

The same goes for pump- and lever-action guns. Jack a round into the chamber and slip another into the magazine. Plus one, it's like magic. However, I'm afraid your top-loading bolt-action is out of luck. The bolt must be retracted to load the magazine.

Having one more round is always nice when you're heading into a fight.

Press Check

The press check is a great way to see if you have a round chambered without unloading it. All you have to do is draw back slightly on the action and see if there's a round in the chamber. This can be done with rifles and pistols.

Some firearms have an indicator that shows whether a round is chambered. This is a handy feature that removes the need to press check.

The press check is an instant credibility builder with firearm aficionados. I always appreciate it when I see it on the screen. However, if you're going to use it in your writing, you'll need to educate the average reader as to what's going on.

Order of Load

This is for characters with serious training. There's an order to loading weapons, which trained characters should always follow when gearing up to go out.

1. Tertiary – Breeching guns and specialty weapons
2. Secondary – Sidearms
3. Primary – The rifle

It's easy to lose track of what's been loaded and what hasn't, especially when you've got a lot of gear. Load your weapons from least likely to use to most likely.

My loadout ritual went something like this:

1. Check all magazines.
2. Strap on the pistol and press check.
3. Put on body armor.
4. Ensure the shotgun has an empty chamber, the action is released, and the magazine is full. (See shotgun breach on page 116)
5. Sling the shotgun.
6. Press check the pistol again
7. Load or press check the rifle. Make sure it's on safe and sling it.
8. Put on my helmet.
9. Perform a radio check.

You'll note that I press check the pistol twice. Anytime I strap on a piece of equipment, I check it. I check it again to follow the procedure. Old habits die hard.

Counting Rounds

In fiction, I've seen a character count the number of rounds their opponent has fired. This may work in westerns, where everyone has a six-shooter, but in a world full of semi-autos keeping a count of the number of times your opponent has shot at you is a near impossibility.

I've also seen a character keep a count of how many times they've fired. In my experience, this is also pure fiction. I'm not saying it can't be done, but I think it's unlikely. I've never been able to do it. There was always too much going on for me to count my shots. In my mind, I either shot a little or shot a lot. To solve the uncertainty of knowing how much ammo was in the gun, I would take a simple action, the tactical reload.

Firing on Full Auto

I find firing on full auto to be imprecise and wasteful. Fully automatic fire should be reserved for suppressive fire – keeping the adversaries' heads down, and effectively engaging moving vehicles. If it's to be done, it should be done in short bursts that allow the character to control the gun.

Tactical Reload

The tactical reload is nothing more than swapping out a partial magazine for a full one. If your character has been in a firefight, they're probably concerned with observing what their opponent is doing and countering their moves. They may move from one piece of cover to the next, relay information and instructions to teammates, and perform a whole host of other activities. If they've lost track of how many rounds they've fired, they can simply change magazines and know that they're full.

This is a good practice to get into, as any *Fortnite* player will tell you. However, in the game, you see your avatar discard the partial magazine. Oof, don't do that. Keep those partial mags. You never know when you'll need them.

Tactical reloads can also be performed on lever- and pump-action guns. Keep feeding rounds through the loading port, if the gun is equipped with one. Bolt action guns can also plus back up, but only if the bolt is retracted.

Interlude: Bump Stocks and Trigger Mods

I've already expressed my opinion on full-auto fire, or spray and pray. Spray a bunch of bullets and pray that you hit something. But some folks just need to feel that recoil.

I also talked about how certain gun enthusiasts attempt to circumvent restrictions on short-barreled rifles. Bump stocks and some after-market trigger assemblies work in a similar fashion only with full-auto firing.

Bump stocks use a clever mechanism that allows the semi-auto rifle to mimic full-auto firing.

Upon firing, the receiver moves back slightly on the stock. This resets the trigger without the need for the shooter to move their finger.

Bump Stock

The receiver is then driven forward by a spring in the stock. The trigger meets the stationary trigger finger, and the rifle fires again.

There are rapid fire trigger mechanisms, like the Hellfire trigger system, that accomplish the same task.

The end result is the same, a semi-automatic rifle converted to fire in a "simulated" fully-automatic mode. Gun enthusiasts and lobby groups argue that these devices fit within the letter of the law. The BATF feels otherwise. The legality of these devices has yet to be decided at the time of this writing.

If you decide to give your character one of these devices, they will, in essence, have a fully automatic firearm. They'll burn through ammunition at an astonishing rate if they don't moderate their fire.

Also, as noted in the machine gun section, the barrel will heat up quickly. If the rate of fire is sustained at a high level, there will eventually be damage to the bore, which will result in decreased accuracy and possibly permanent damage to the barrel over time.

Engagements should be limited to short bursts, no more than three to five rounds at a time. At this rate, your character will burn through a thirty-round magazine in six to ten bursts.

Drum magazines can hold up to one hundred rounds, but they make the gun pretty heavy. This is actually a good thing as the added weight helps manage recoil.

Handgun Overview

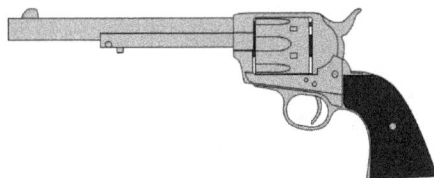

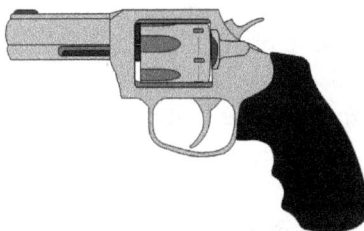

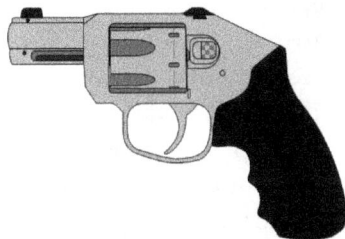

Just as long arms went through evolutionary steps, so did handguns. Since the first pistols mirror the musket, there is no need to cover them here. Your characters will take the same actions to load and fire single-shot, muzzle-loading pistols as they would with muskets.

Before we get started, I must make a distinction between handgun types:

Pistol

- Pistol: A handheld firearm with one or more firing chambers.
- Revolver: A handheld firearm with multiple firing chambers housed in a rotating cylinder.

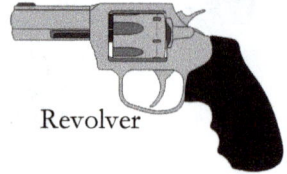

Revolver

Pistols and revolvers are not the same thing, even though several dictionary sites don't make a distinction. Even the ATF website defines a pistol as, "a weapon originally designed, made, and intended to fire a projectile (bullet) from one or more barrels when held in one hand."

Hell, given the ATF's definition, a rifle could be considered a pistol if it were fired one-handed. It'd be awkward but it could be done. Having said that, check out the interlude for Rifle Caliber Pistols. (Page 107)

Anyway, these two forms of the handgun are not the same animal. Just like cats and dogs can be pets, but they're not the same type of creature. Is it a great sin to use the word *pistol* for both? No. The dictionaries of the world will back you up. However, being precise will remove reader confusion and possibly save you a bad review from some pedantic knucklehead.

There are plenty of generic synonyms out there to give your writing variety without confusing the type of handgun.

The oddballs, derringers, and other multi-barreled, multi-chambered handguns fall into the pistol category.

We're not even going to talk about the harmonica gun, but it would be classified as a pistol.

COP .357
4 barreled
Deringer

Pistols and revolvers are further classified by trigger action:

Single-Action (SA)

The trigger only performs the action of releasing the firing mechanism to fire the gun.

For revolvers, this means cocking the hammer every time the shooter wants to fire. When the shooting is over, the character simply lowers the hammer, or sets it to half cock, to make the weapon safe. Single-action revolvers are archaic. Feature them only through the 1890s.

Single action Revolver

In the semi-autos, a single-action trigger doesn't require the shooter to cock the firing mechanism for each shot. The pistol does that for them when it

Single action Pistol (1911)

cycles. Likewise, the character should never lower the hammer on a semi-auto, single-action trigger. Instead, they should engage the safety. More on this later.

You may have noticed that the definition uses firing mechanism rather than hammer. Single-action revolvers will use a hammer, but single-action semi-autos might not.

The 1911 pictured above uses a hammer. Other semi-autos may use what's called a striker mechanism. In guns of this type, there is no hammer for the operator to manipulate. The firing pin is renamed "striker," and it's held in place by a sear. The trigger only performs the action of releasing the sear.

When the pistol is initially loaded, the striker is set under spring tension. When the gun is fired, the striker is reset by the action of the gun.

For our purposes, because there's no hammer, we'll treat all striker-fired pistols as single action.

WAIT! Don't rush off and write that in a pistol's description. We'll all get in trouble for that. What I mean is that when your character interacts with the gun, the only thing they can do with it is load, fire, unload, and put the gun on safe … if it's equipped with a manipulatable safety.

Double-Action/Single-Action (DA/SA)

In pistols of this type, the trigger performs both the action of cocking the hammer and releasing it to fire the gun. If cocked manually, or by the action of the pistol, the trigger only performs the action of releasing the hammer. The double-action (DA) trigger is considered a safety feature because it requires a long, deliberate trigger pull.

In both revolvers and pistols, mechanisms inside the gun draw back the hammer as the trigger is pulled. When the trigger reaches the break point, the hammer is released and the gun fires.

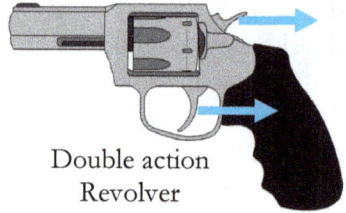

Double action
Revolver

If your character manually cocks the hammer, the trigger is set to full cock and it becomes a single-action trigger.

The single-action position requires less force to be applied to the trigger for the gun to be fired. Also, the trigger doesn't need to travel the same distance to reach the break point.

Single action when
cocked

In revolvers, the cylinder is rotated to a fresh chamber as the hammer is drawn back.

In semi-autos of this type, only the first shot is double action. The hammer is cocked by the action after every shot. The hammer is lowered to a safe position either by the safety or a de-cocking mechanism.

Double action
Pistol

Single action
after 1st shot

This is my favorite trigger type for writers because there are so many possible ways for the character to interact with the pistol.

You can escalate:

She drew back the hammer ...

Or de-escalate:

She de-cocked the Beretta ...

Striker-Fired

The trigger interacts with a firing mechanism through a variety of actions. If there's no external hammer, it is irrelevant how the striker is manipulated. However, the manufacturer's description of the trigger action can be confusing and lead you to make mistakes. These descriptions include single-action, double-action, or safe-action.

Striker fired pistol
Glock 19

In safe-action pistols, as the trigger is pulled, multiple safeties are disengaged, the firing pin is retracted, then released at the break point.

Double-Action-Only (DAO)

The trigger performs both the action of setting the firing mechanism and releasing it. There is no way to manipulate the hammer.

From a writer's point of view, I lump SAO, DAO, and striker-fired pistols all into the same category: they're about as dramatic as a brick.

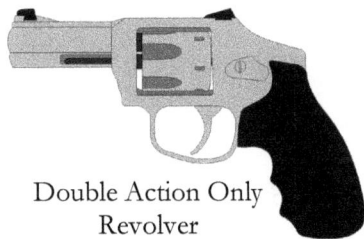

Double Action Only
Revolver

Everything that the gun does is internal. The character has no control over these processes. Character interaction is limited to load, fire, unload, and safe (if an external safety is equipped).

These handguns will have clean lines. Revolvers will appear to be hammerless, and pistols will have a clean back end.

Single-action and striker-fired pistols will generally have lighter triggers, while DAO triggers are usually harder to pull and have a longer travel distance.

Summing up: handguns can be classified as either pistols or revolvers and are further defined by their trigger type, or action.

I'll provide a few historical examples of revolver and pistol design, then we'll start getting into the fun stuff.

Derringers and Revolvers

1825-Present

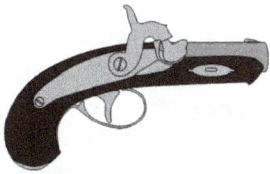 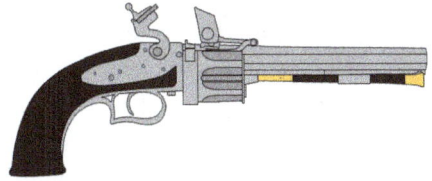

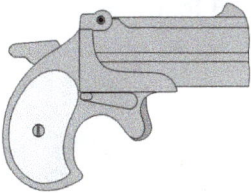 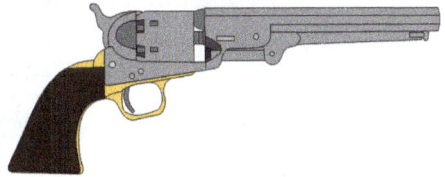

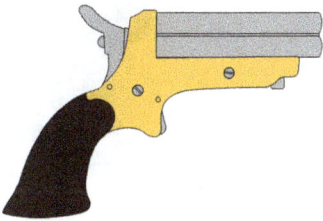 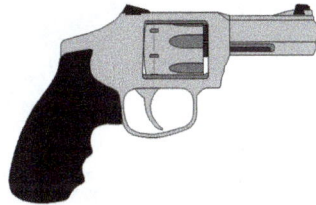

Derringers
Philadelphia Deringer, 1825

The first derringer was developed by Henry Deringer. It was a single-shot, muzzle-loading, caplock pistol small enough to fit in a pocket. The design was so popular that Deringer sold fifteen thousand of these guns, and his name became a household word associated with any pocket pistol.

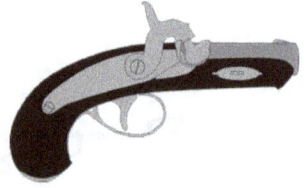

The spelling and capitalization of the word has varied over the years. *Deringer*, after the man, was the original spelling. In the late 1850s, misspellings added another R to the name, *Derringer*, which remained capitalized. Lowercase versions of both spellings soon followed as the derringer was considered a class of pistol. Today, *derringer* is the recognized spelling of the word.

While Colt released a single-shot derringer in 1870, most are recognized as having multiple barrels. It's these we'll focus on.

Sharps Derringer, 1859

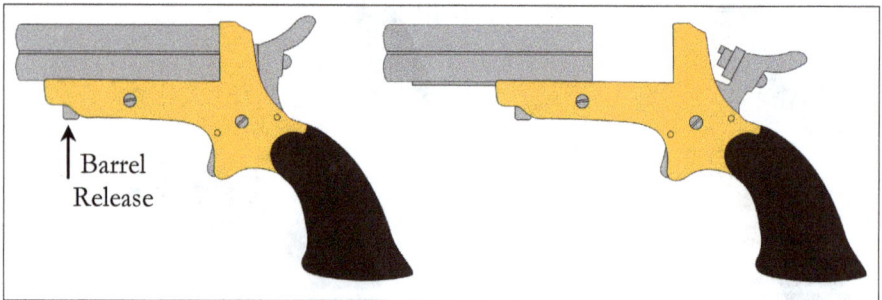

Barrel Release

The Sharp's pepperbox was a four-barreled derringer that had a rotating firing pin on the hammer.

The action was opened by depressing the button on the bottom of the gun and sliding the barrel assembly forward.

The four-barrel design was not an evolutionary dead end, although it did go dormant for a while. COP resurrected the design in 1983 and Advantage Arms had a similar model out in 1986, but these guns are no longer produced.

Current manufacturers of the four-barrel derringer include Signal 9 Defense and Iver Johnson.

The derringer finds its recognized form in the two-barrel design pioneered by Remington. The four- and three-barreled versions needed to reduce the size of the round to make them more concealable.

While certainly deadly, the smaller rounds didn't provide the definitive power of a larger caliber.

138

The double-barrel derringer has been manufactured by various companies up to the modern day and almost all of them follow the Remington design.

Remington Derringer, 1866

As with all derringers, this design also features a single-action trigger.

To load, the shooter rotates the barrel release lever all the way forward. This releases the barrel assembly to rotate upward.

New rounds are inserted, and the barrels are rotated back into position and locked into place.

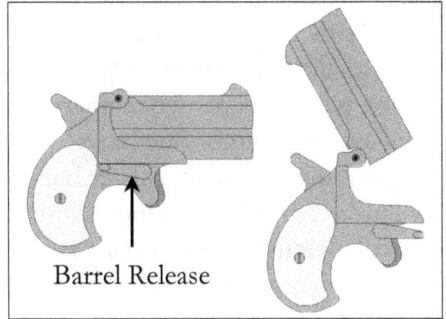

Barrel Release

Newer designs of this pistol may incorporate features you'd expect in modern firearms.

- Action Release Lever: Newer models put the lever under spring tension, relieving the need to rotate the lever back into place.
- Trigger Guard: This is a removable option in some models. The shooter can decide to retain or remove it at their discretion.
- Positive Safety: The safety will be in the form of a button on the side of the frame near the hammer.
- Ejector **Tab:** The ejector tab slides backward to eject spent casings or clear the gun.

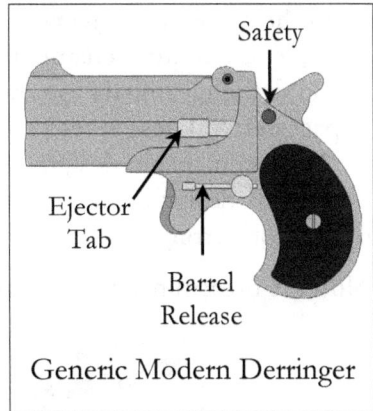

Safety

Ejector Tab

Barrel Release

Generic Modern Derringer

Interlude: The Pepperbox

A pepperbox is a multi-barrel (three or more), multi-chamber firearm, usually operated in a revolving manner, with a single trigger mechanism. This design sacrifices weight and simplicity in favor of safety for the shooter.

Early multi-chamber handguns were prone to chain fire, which occurs when sparks from the firing of one barrel ignite the charge in a sister barrel. This occurred in flintlocks and early percussion cap firearms.

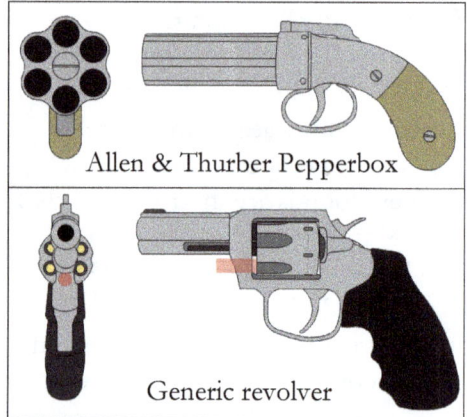

Allen & Thurber Pepperbox

Generic revolver

With a pepperbox gun, there are no parts of the frame to get in the way of a sympathetically fired chamber or barrel.

In the modern revolver example, the bottom chamber would definitely impact the frame. The other chambers represent a frame impact risk with less severe consequences.

Advancements in design and materials removed the need for the pepperbox design in larger-sized handguns. However, it remains a good option for derringer pistols.

Note: A firearm needs three or more barrels to qualify as a pepperbox.

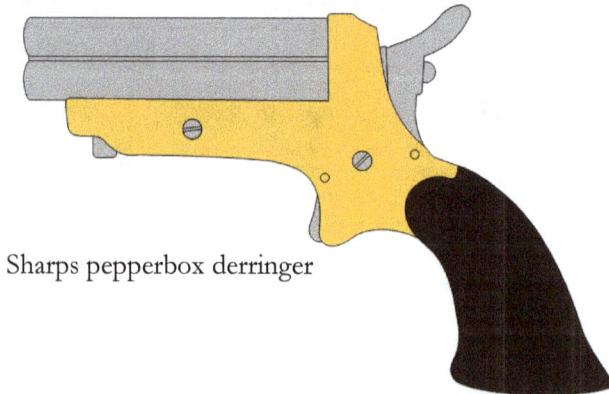

Sharps pepperbox derringer

Revolvers

I know I said we were done with muzzleloaders, and technically that's true. Early revolvers didn't load at the muzzle; they loaded at the chamber.

The Collier Flintlock Revolver, c1800

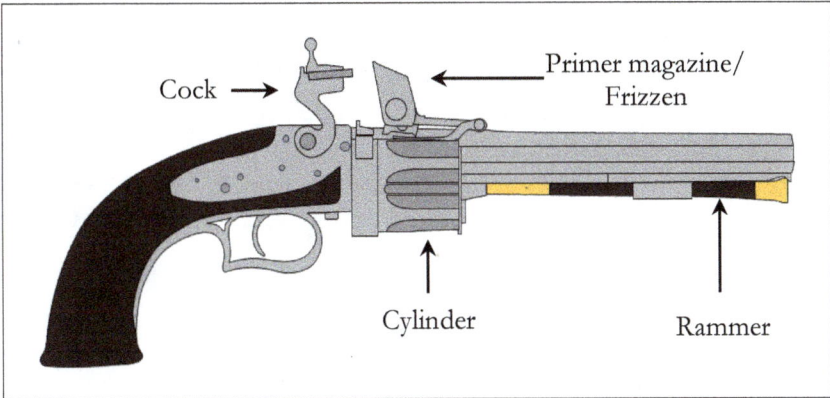

One of the first revolvers was introduced to the world by Elisha Collier in the early 1800s. Each chamber in the cylinder was loaded in the traditional, muzzle-loading way.

After each shot, the shooter needed to cock the hammer and rotate the cylinder by hand. The weapon was equipped with an ingenious priming magazine that primed the pan when snapped into place. This magazine also acted as the frizzen.

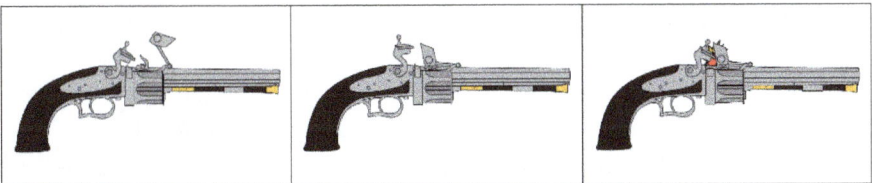

To fire following chambers:
- Set the hammer to half cock.
- Rotate the cylinder manually.
- Rotate the primer magazine/frizzen into place, at which point the primer powder is automatically released.
- Bring the hammer to full cock.

When the trigger is pulled, the hammer swings forward like other flintlocks. Sparks ignite the primer powder, the gun fires, and the primer mag/frizzen swings clear.

Video: https://www.youtube.com/watch?v=Pfm06EcBtcc

The Collier was a great advancement in revolver technology, but it was to be a short-lived success. As with long arms, the invention of the percussion cap revolutionized handgun design. However, these early guns solved many of the mechanical issues for future gunsmiths.

To be sure, there were other pepperbox guns that took advantage of the percussion cap prior to the Allen and Thurber, but this model incorporates rotating the barrel as part of trigger operation. This gun was also extremely popular in its day; thus, it's included here as a representative of its class.

Allen and Thurber Pepperbox, 1837

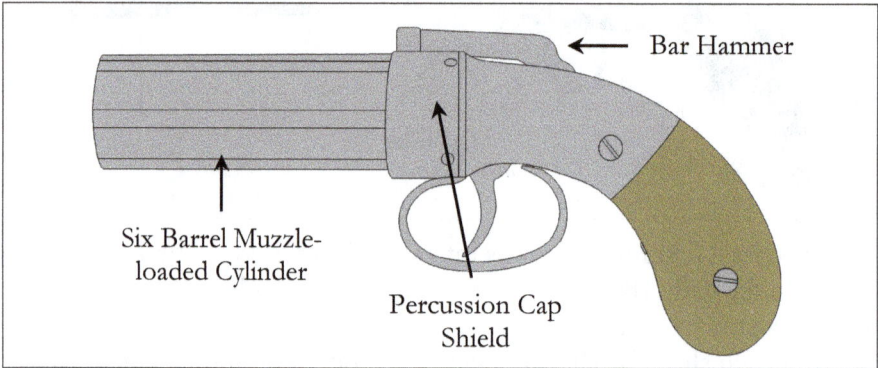

Bar Hammer

Six Barrel Muzzle-loaded Cylinder

Percussion Cap Shield

The six independent barrels were loaded in the same fashion as all muzzleloaders. Percussion caps were then set on cones for each barrel. These aren't pictured in the illustration because they're hidden by the percussion cap shield, which was an attempt to minimize the possibility of chain fire.

The Allen and Thurber also takes a step forward in that it's a true double-action trigger. It harnesses trigger pull to rotate the cylinder. When the trigger is pulled, the cylinder is rotated, and the hammer is raised to the break point, at which time it's released, and the gun fires.

Video: https://www.youtube.com/watch?v=wjReSGFtUtY

Our final entrant in the muzzle-/chamber-loading category is the Colt Navy revolver. This firearm is the final "cap and ball" gun. There are other contenders for inclusion here to be sure, but the Colt is where we see the revolver take its modern form for the first time, albeit as a single-action, chamber loader.

Navy Colt, 1851 (SA)

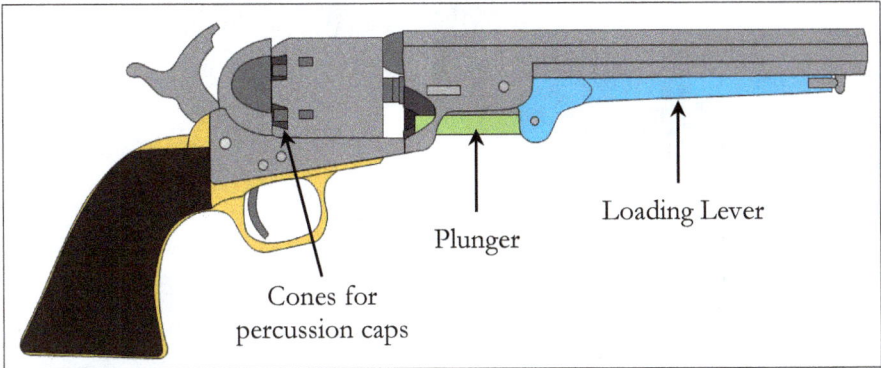

The chambers of the Navy Colt are loaded in the usual fashion, but instead of using a rammer, the Colt uses an integral loading lever to seat the projectile.

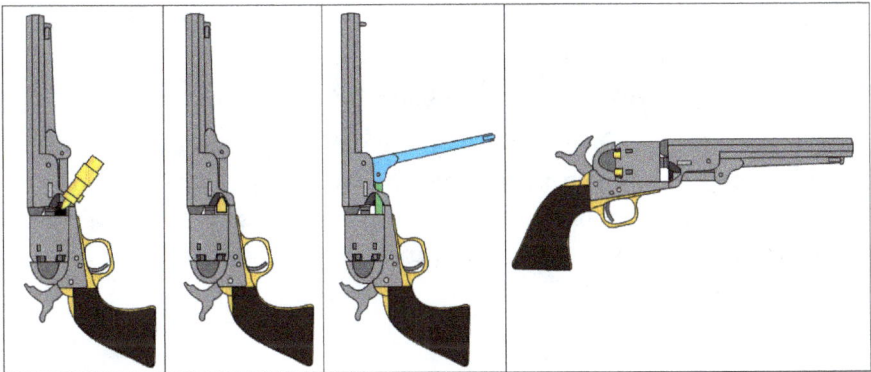

To Load:
- Set the hammer to half cock.
- Pour a measured amount of powder into the chamber.
- Insert the round.
- Rotate the cylinder (by hand) to line up with the plunger.
- Pull the loading lever down so the plunger seats the round.
- Set caps on the cone for each barrel.

Video: https://www.youtube.com/watch?v=2L30-McaZCA

Revolver designs exploded in the late 1800s with the invention of the modern cartridge. All manner of mechanisms were invented, tested, and produced. Here are the highlights.

Lefaucheux M1854, (SA)

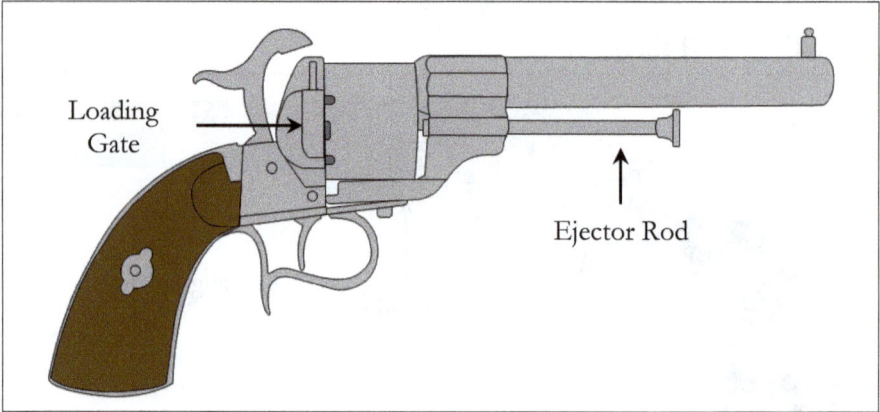

The Lefaucheux is the first revolver to use a metal cartridge. Apart from hammer design, this pistol is the standard for single-action revolvers of the era.

To Load:
- Draw back the hammer to the load position. This frees the cylinder to be rotated by hand.
- Insert rounds individually through the loading gate.

Unloading was performed through the same gate by means of pressing back on the ejector rod. This also ejected spent casings. This rod wasn't under spring tension and had to be manually slid back into place.

The Lefaucheux used the pinfire cartridge. (Page 61)

Tranter, 1863 (DA/SA)

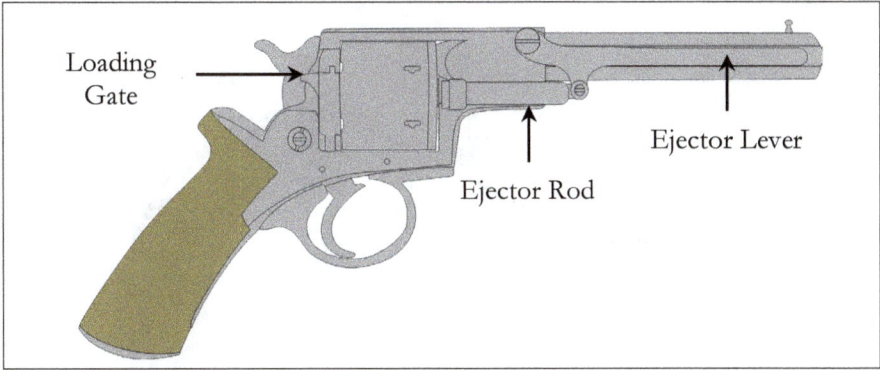

The Tranter breech-loading revolver was the first double-action revolver to use the modern cartridge. It used the loading gate design and a lever-action plunger to clear spent casings.

Colt Peacemaker, 1873 (SA)

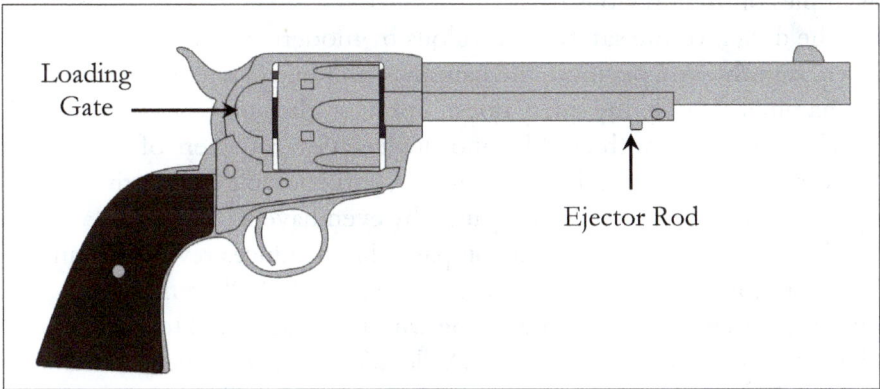

Colt improved on the basic design of the Lefaucheux. First, by putting the ejector rod under spring tension, the shooter didn't need to manually move it back into place.

Apart from convenience, this also removed the possibility the rod could foul the rotation of the barrel. The ejector rod slotted into place when not in use.

Gate-loading pistols enjoyed a good thirty-year run. However, new designs would soon overtake the old system. The first top-break revolver appears to be the Schofield patent Smith and Wesson Model 3, 1870.

The Model 3 was adopted by the US military, but ammunition confusion led to most servicemen retaining their Colt .45s.

Reichs Revolver, 1879 (DA/SA)

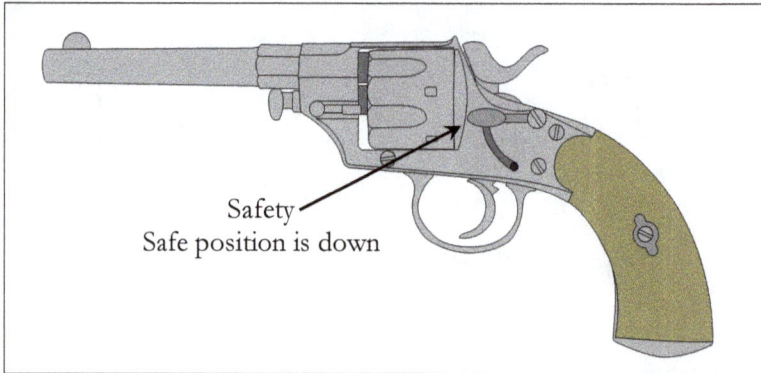

Safety
Safe position is down

The Reichs revolver is the representative of all revolvers equipped with safeties produced in the late 1800s and early 1900s. Revolvers with this feature were mainly produced in Europe.

These revolvers are all loaded through a loading gate similar to the examples on the previous page.

The design of the safety is ridiculous by modern standards. When the safety is in the safe position, the hammer can't be cocked. However, if the hammer is back, the safety doesn't prevent the gun from firing!

This is true for both double- and single-action revolvers of this type in this time period. In fact, in some models, you couldn't engage the safety if the hammer was back. Ridiculous, why even have it?

Heritage Manufacturing is a company that produces revolvers with a hammer block safety. When the safety is engaged, the hammer is physically prevented from interacting with the firing pin. Heritage only manufactures recreations of classic Wild West–style revolvers. Their product list is currently limited to single-action .22-caliber guns.

Videos:
General Video:
https://www.youtube.com/watch?v=NWwzyBRFQMM

Review of a Heritage Revolver:
https://www.youtube.com/watch?v=c6y-KVopb-o

Webley Mark 1, 1887 (DA/SA)

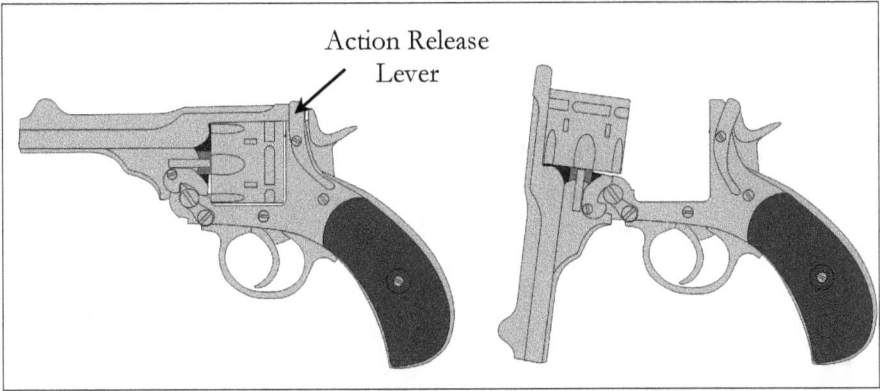

Action Release
Lever

Meanwhile, across the pond. The Webley-Price revolver was the first British revolver to see the top-break design. Webley would score big with his Mark 1 design (pictured here). The Mark 1 remained in service until 1970.

The action is opened by a lever on the left side of the frame. The barrel and cylinder assembly tilted forward, which gave the shooter access to all of the chambers at once, instead of having to eject spent casings individually.

Smith & Wesson Model 1 Safety Revolver, 1887

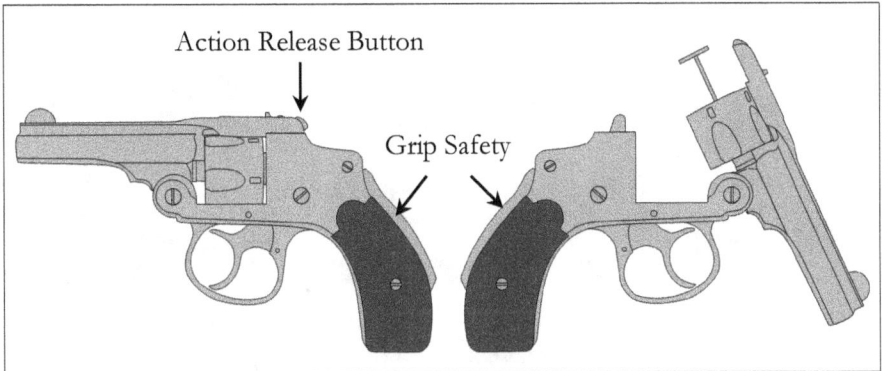

Action Release Button

Grip Safety

The Model 1 features the top-break design. It also gets the prize for being the first to create a hammerless revolver. These guns are considered double-action-only in modern terminology.

Another interesting feature of this revolver is the grip safety. It can't be fired unless the grip safety is depressed.

The Modern Revolver

Eventually, Colt designed the first swing out cylinder in 1889. The cylinder is released by manipulating the cylinder release lever. The cylinder swings out, and the shooter uses the ejector rod to push out empty cartridges.

All modern revolvers use the swing out cylinder and ejector rod reloading system. There hasn't been a significant change to loading operations since the adoption of this design.

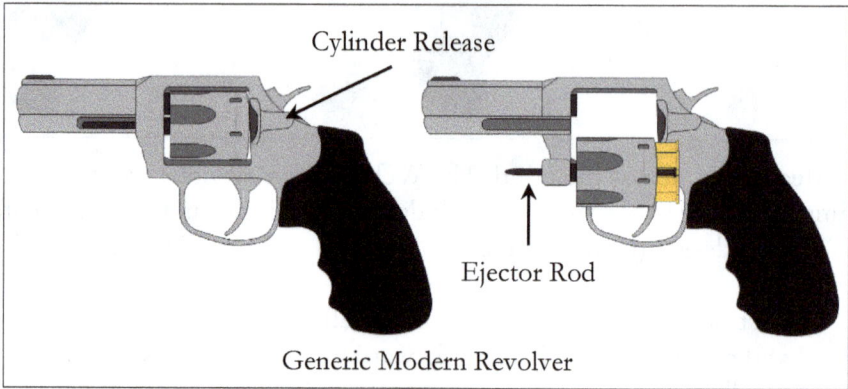

Cylinder Release

Ejector Rod

Generic Modern Revolver

The main design differences are in how the cylinder release is activated. The release mechanism is almost always on the left side of the gun and can be one of three types.

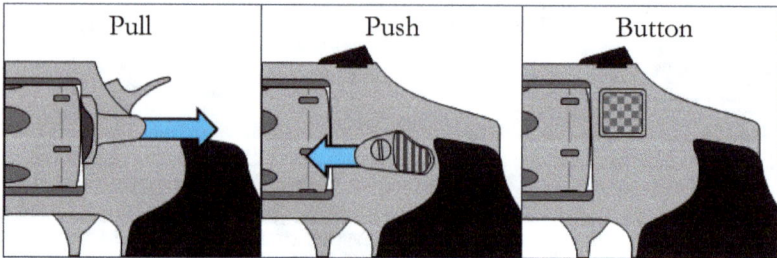

Pull   Push   Button

- Pull: The release mechanism is pulled toward the rear of the gun. Pull type will generally have a knob that the finger or thumb can manipulate easily.
- Push: The release mechanism is pushed toward the front of the gun. Push type will generally be textured to prevent the finger or thumb from sliding off.
- Button: The release mechanism is a button that must be pushed in to release the cylinder

Regardless of type, force must be applied to the cylinder to rotate it out of the frame.

Modern Revolver Variants

Modern revolvers are available in two different configurations.

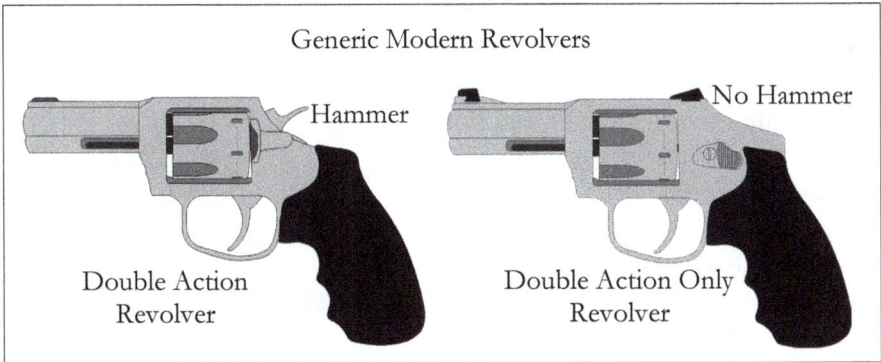

Generic Modern Revolvers

Double Action Revolver — Hammer

Double Action Only Revolver — No Hammer

Double-Action

Double-action revolvers are equipped with a hammer that may be cocked by the shooter. This allows the gun to be fired in either single- or double-action mode.

Double-Action-Only

DAO revolvers have no hammer that the shooter can manipulate. The trigger performs the expected action of rotating the cylinder. As it rotates, it cocks and releases the firing pin. If the trigger is released before firing, the firing pin returns to a state of rest and is no longer under spring tension.

Revolver Capacity

The design of the cylinder depends on the number of chambers and caliber of ammunition.

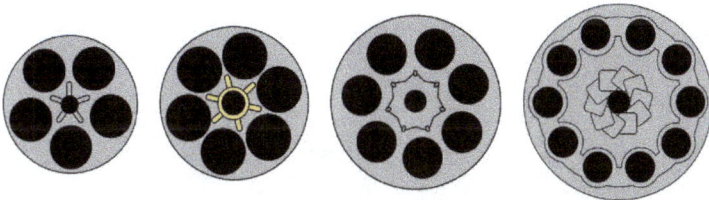

Revolvers that fire large caliber rounds generally contain fewer chambers in the cylinder. However, if the diameter is increased, the number of chambers may also be increased.

Chamber count commonly runs from five to ten and hits every number in between. However, some of the older revolvers had a ridiculous number of chambers.

Don't assume that all revolvers are six-shooters. Always check the caliber and capacity when choosing any firearm.

Fluting

The grooves in the cylinder are called fluting. Fluting isn't decorative. It serves two purposes. The first is weight reduction. The rounded shape allows manufacturers to reduce the amount of metal without reducing the ability of the chambers to handle the pressure of firing.

Fluting also makes rotating the barrel easier. This was an important feature in the single-action revolvers where each chamber had to be loaded individually.

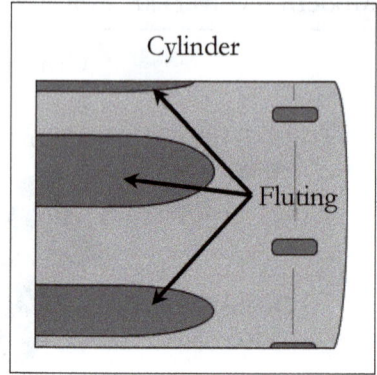

Cylinder

Fluting

Video:

De-Cocking the Revolver:
https://www.youtube.com/watch?v=O8-cbgzeQSY

Single-Action Revolver Operation

The following is a general description of how the single-action revolver operates. If you're going to use a single-action firearm other than the Colt Peacemaker in the following example, confirm the details of its operation.

Single-action revolvers have multi-position hammers for safe operation. This is evidenced by the number of clicks you hear as the hammer is pulled back. There are generally three main positions.

Safe Position

The first click back is the safe position. When the hammer is drawn back, the integral firing pin is not resting on the cartridge. In single-action revolvers with a hammer and firing pin design, the hammer is not resting on the pin.

The second click back is the loading position, or half cock. This is a common position for all the chamber and gate-loading revolvers. With the hammer in this position, the cylinder is released and is allowed to rotate freely. This allowed the shooter to push out spent cartridges and load in fresh ones.

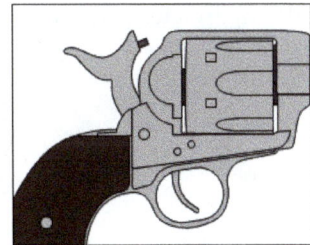

Half-cock

To move between the safe and half-cock positions, the hammer must be retracted slightly to disengage it from its lock notch. (See the drawing for half cock in the lock interlude on Page 42.)

The third position is full cock, ready to fire. The hammer must be manually positioned to full cock for every shot.

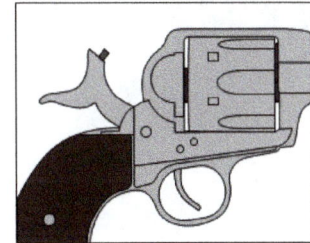

Full-cock

Quick-Draw Techniques

There are two methods to fire a quick draw.

Thumbing: As the shooter draws the gun, they use the thumb of the firing hand to draw back the hammer to full cock and then they pull the trigger.

Fanning: The shooter applies pressure to the trigger as they draw the gun from the holster. Once they feel the gun is lined up with the target, they bring the other hand in a sweeping motion across the hammer, bringing it to full cock. Since the trigger has already been pulled, the hammer doesn't catch in place. It falls forward, firing the gun.

Fanning can also be used to fire many shots in rapid succession. However, it's usually inaccurate and puts undue strain on the gun and its parts.

As an added measure of safety, revolvers of all types are often carried or staged with the first chamber empty. Yes, it reduces the capacity of the gun, but it also prevents a misfire when the shooter lowers the hammer to the safe or fully down position.

Video:

Quick-Draw Techniques:
https://www.youtube.com/watch?v=wr0azVhRZUQ

Double-Action Revolver Operation

The following is a general description of how post-1900 double-action revolvers operate. If you're going to use a double-action revolver that existed prior to the 1920s, confirm the details of its operation.

Double-action revolvers are designed to have either a slight standoff between the hammer and the firing pin or a firing pin block when the hammer is in the down position and the trigger is forward.

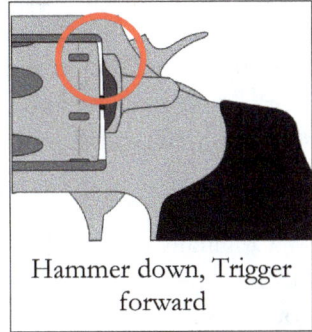

Hammer down, Trigger forward

This makes the gun safe to carry with the hammer down and a fully loaded cylinder. The double-action revolver is an inherently safer firearm than its single-action brethren.

When the trigger is pulled, the cylinder rotates as the hammer comes back, the safety mechanism that creates the standoff is rotated out of position, allowing the hammer to fall on the firing pin which, in turn, strikes the cartridge as shown in the illustration.

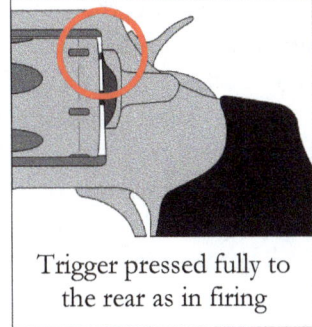

Trigger pressed fully to the rear as in firing

When the trigger is released, the safety mechanism rotates back into place, re-establishing the standoff or firing pin block.

De-Cocking

It's important to remember the safety mechanism when de-cocking double-action revolvers. Many people keep the trigger fully depressed when de-cocking. This can lead to an accidental discharge.

The proper way to de-cock is to block or hold the hammer, then pull the trigger. Once the hammer is released, allow the trigger to go forward to engage the safety mechanism, then lower the hammer all the way down.

Interlude: Speed Loaders

The revolver was a great leap forward, but it had one major drawback: it was still slow to load. Sure, the modern cartridge sped things up, but ejecting and loading cartridges through the loading gate was laborious work. With the advent of the top break and swing-open cylinders, shooters could reload much faster. If only there were a tool for this ...

Introducing the speed loader!

Originally designed in 1879 by William H. Bell, these handy devices make loading your revolver a breeze.

Joking aside, the speed loader is a great invention, but there are still a couple of drawbacks.

Speed loaders must be of the same configuration as the cylinder of your character's revolver. If the chamber count or spacing is different, it won't work.

There are many designs and release options available. If your character uses them, make sure you know the type and proper release action.

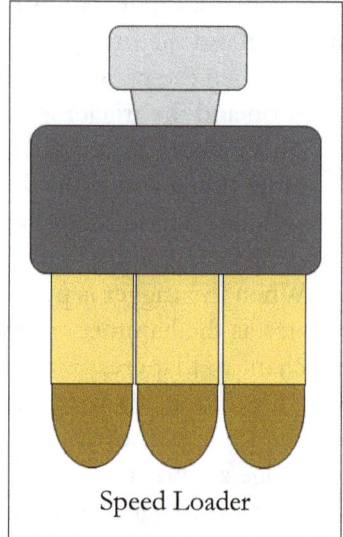

Speed Loader

Another option for revolver users is the speed strip. These handy devices are metal, polyurethane, or neoprene strips that the rounds fit into. They aren't as bulky as the speed loader, but they are a bit slower to use.

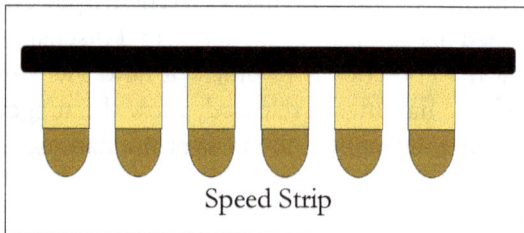

Speed Strip

To use the speed strip, insert the nose of the cartridge into the cylinder, then pull it free of the clip.

Webley-Fosbery, 1895

As we head into the automatics, I couldn't resist one oddball – a truly innovative revolver that came to the automatic class just a little too late to be relevant. However, since it's mentioned in the *Maltese Falcon*, it bears inclusion here.

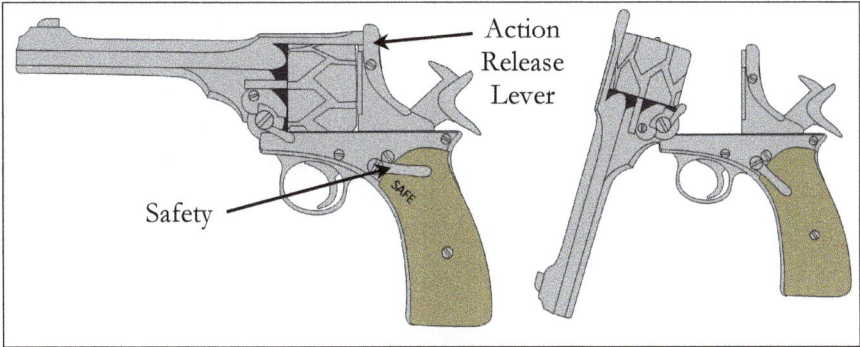

The Webley-Fosbery is the only semi-automatic revolver ever created. It had a single-action trigger and a positive safety. Unlike the other revolvers with safety levers, the safety on the Fosbery prevented the gun from firing with the hammer back. The gun could be carried safely in a cocked and locked condition.

Like the Webley Mark 1, the Fosbery had a break-action cylinder that allowed for quick reloads.

Operation

When the Fosbery fired, the entire top half of the gun was forced backward by recoil, cocking the hammer. At the same time, grooves in the cylinder interacted with a stud in the frame, rotating the cylinder, and bringing the next chamber in line with the barrel.

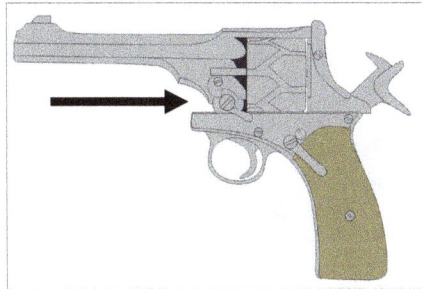

Less than five thousand of these guns were ever produced, so I can't see how Sam Spade could toss off its mention so casually in the book. Seems to me that it would be an important plot point. Then again, Mr. Hammett did like his red herrings.

Automatic Pistols

1892-Present

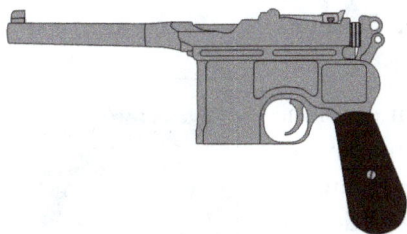

While revolver designers were going nuts with their designs, the evil geniuses on the automatic side of the aisle had to wait for the invention of smokeless powder to get their designs off the ground. Once that genie was out of the bottle, the family of automatic weapons was born.

As you may expect, the first designs were kind of crazy. Some inventors used the trapdoor approach. Take an existing design, retool certain parts, add a spring here, a lever there, and presto! Automatic pistol. Others designed their automatics from the ground up.

The first patents for automatic pistols start to appear in the early 1890s. It's tempting to include the progenitors here in this work, but most of these guns, no matter how influential, rarely exceeded trial runs. Here's a quick rundown of the top three finishers.

Salvator-Dormus Pistol, 1892

This pistol takes first place in the patent division. It beat out the second-place contender by eleven months. About fifty of these guns were produced and submitted to the Austrian military as a design for their service pistol. It was deemed inferior and rejected.

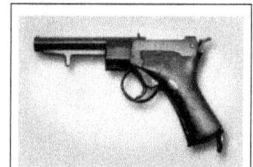

Salvator-Dormus Pistol -Torana

Schönberger-Laumann, 1892

This pistol gets the prize for second place in the patent division. Sadly, only thirty-five of these guns were ever made. These prototypes were also submitted to the Austrian military, and they too were denied.

Borchardt C93, 1893

This pistol takes third place. The Borchardt experienced limited success; about three thousand of these guns were manufactured. A number of countries performed trials on the Borchardt, including the US

Borchardt C93 -Mike Johnstown

Historically significant? Yes. Relevant to fiction? Not really. Most characters will use a revolver prior to 1896. However, your late nineteenth-century spy could reasonably use one of these early designs. The Dormus is the most advanced, but the Borchardt is more common.

Interlude: Semi-Automatic Pistol Firing Cycle

This basic cycle applies to almost all semi-autos. Again, this is an oversimplified drawing.

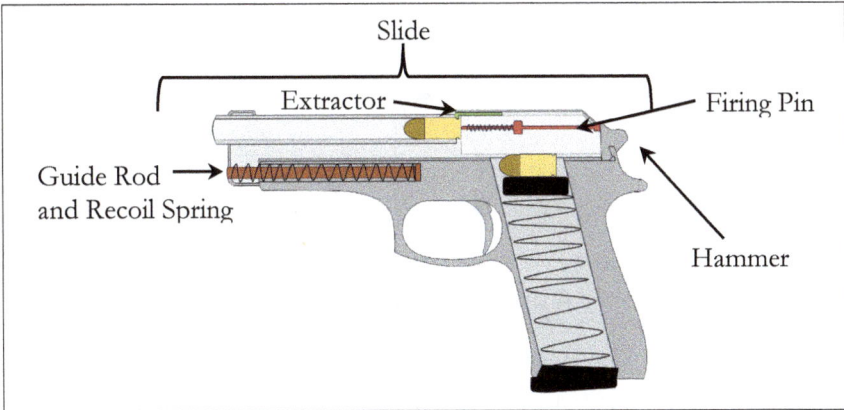

When the trigger is pulled, the hammer is released and it swings forward into the firing pin, which fires the round.

Blowback from released energy drives the slide backward, and the extractor pulls the expended casing from the chamber and compresses the recoil spring.

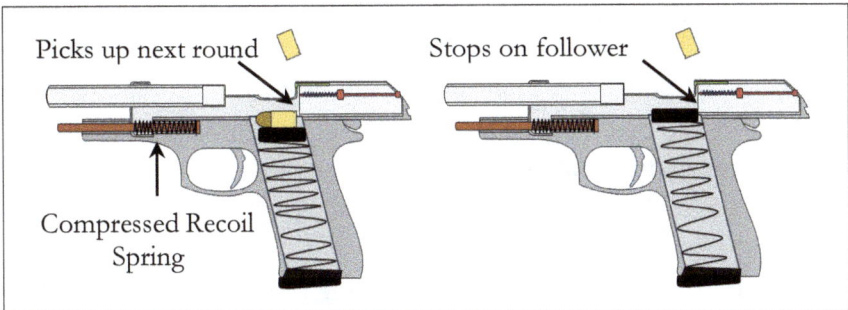

As the slide reaches the length of its travel, the trigger is disconnected – to prevent automatic fire, and the expended casing is ejected.

As the slide is driven forward by the recoil spring, it picks up the next round from the magazine and loads it into the chamber.

The trigger is reset when the shooter releases it.

When the last round is fired, the slide catches on the follower in the magazine halting its progress. This allows the slide release to engage the slide and halt the cycle. With the slide back, the trigger is disconnected, and the gun won't click when empty.

Note: Not all semi-auto pistols have the slide lock feature. They will click once when empty.

Mauser C96, 1896 (SAO)

The Mauser C96 is arguably the first truly successful automatic pistol. While other automatics were under consideration for adoption by various nations, the C96 was out in the field getting shit done.

Jonathan Ferguson says it best in his book *The 'Broomhandle' Mauser*.

> *Overall, it has to be said that the C96 pales in comparison with later pistol designs and would be unsuitable for today's various military, police and civilian needs. Nevertheless, it must be remembered that in 1896 there was simply nothing in its class to touch its firepower, reliability and accuracy potential.*[2]

The C96 was the first semi-automatic pistol used in combat, and Winston Churchill carried this pistol while he served in the British army.

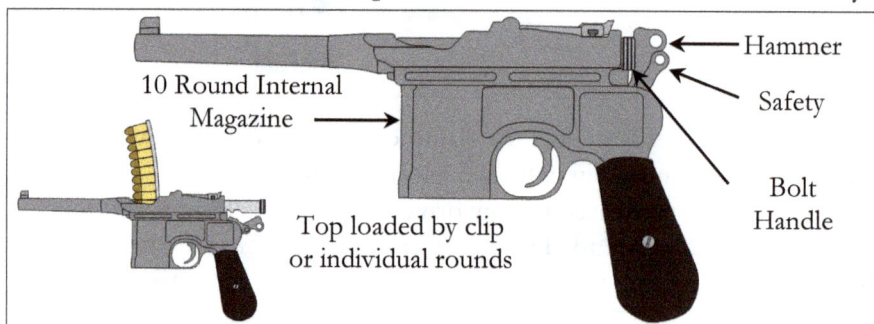

10 Round Internal Magazine →

Top loaded by clip or individual rounds

→ Hammer

Safety

Bolt Handle

To Load:
- Pull back on the charging handle until it locks.
- Insert a clip and press all rounds into the magazine. When the clip is removed, the bolt will go forward.
- Rounds can be loaded singly if the charging handle is held to the rear.

The bolt handle locks to the rear when the last round is fired.

It should be noted that this gun has several variants as the result of general improvements and licensing to various manufacturers. The safe position/engagement procedure is not always the same.

There were over one million of these guns produced between 1896 and 1937. However, with the advent of more efficient pistols, the C96 fell into decline and became a weapon favored by gangsters and revolutionaries.

---

[2] Jonathon Ferguson, *The 'Broomhandle" Mauser, (Weapon Book 58)* (Oxford: Osprey, 2017), Kindle, 76.

Luger P08, 1898 (SAO)

The Luger P08 is the direct descendant of the Borchardt C93. The Borchardt was a revolutionary pistol, but it had issues. Georg Luger was brought in to fix them. Originally designed in 1898, the Luger used the Borchardt's reciprocating toggle action. This means every time the gun is fired, the toggle arms spring up and down as the gun cycles.

It went through various design changes before being adopted by the German military in 1908, Pistole 08.

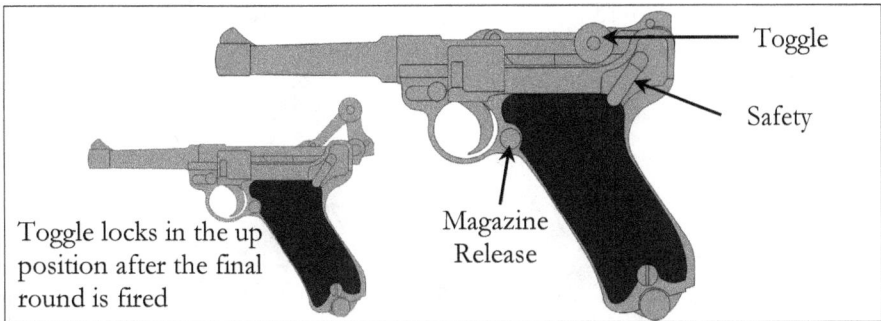

Toggle

Safety

Toggle locks in the up position after the final round is fired

Magazine Release

The Luger's innovative incorporation of the detachable box magazine in the grip has been the standard for almost all modern pistols since. The Luger had one major flaw. It was expensive to produce. Despite this, the P08 remained in service from 1900 through the 1960s. Not bad for a pistol with such a strange action.

There were over three million P08s produced, and the Luger is highly sought by gun collectors. Production of this gun ceased in 1986.

To Load (Two Methods):
Method One
- Insert the magazine
- Draw back and release the toggle assembly

Method Two
- Lock the toggle to the rear
- Insert the magazine
- Release the toggle.

When the last round is fired, the toggle locks in the up position allowing for faster reloads.

To reload, the shooter ejects the spent magazine, inserts a new fresh one, draws back on the toggle and releases, chambering a round.

Nambu Pistols, (SAO) 1904

The Nambu Type 14 was the main service pistol of the Japanese army in WW2. The original design, the Type A, appears to have originated in 1904, which is why I'm covering it here.

There were less than four hundred fifty thousand Nambu variants produced between 1906 and 1945. This is a staggeringly low number for a nation that fought in a world war. The reason is cultural. The Japanese are raised in the culture of bushido. For this reason, the sword, rather than the pistol, was held in higher regard. At the end of WW2, edged weapons outnumbered handguns.[3]

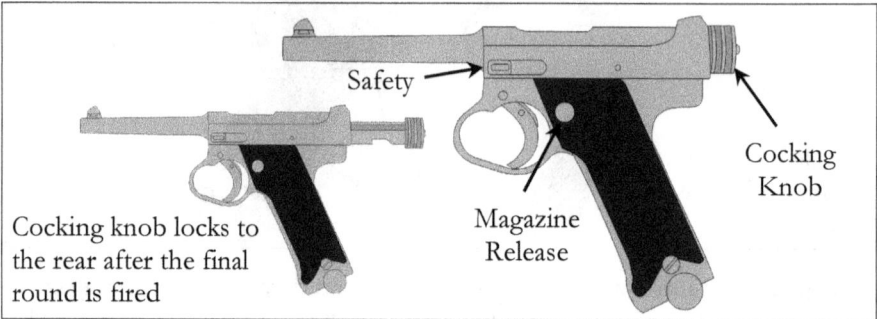

Safety

Cocking Knob

Magazine Release

Cocking knob locks to the rear after the final round is fired

To Load:

- Insert the magazine and draw back the cocking knob.
- When the last round is fired, the bolt locks to the rear.

The Nambu pistols were terrible guns. They were poorly designed and prone to malfunction. Then why include it here? I'm glad you asked.

After the war, a guy named Bill Ruger got ahold of a captured Type 14. He tinkered around with it in his garage and eventually designed the Ruger Standard Model, chambered in .22 LR.

The Ruger Standard pistol would go on to be one of the most successful .22-caliber pistols ever made. It's still in production as the Ruger MK IV.

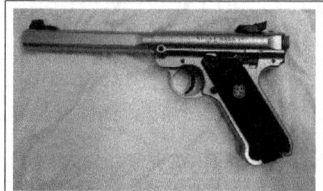

Ruger MK IV

-Coati077

---

[3] John Walter, *Nambu Pistols: Japanese Military Handguns 1900–45 (Weapon Book 86)*, Oxford: Osprey, 2023, Kindle, 34.

Colt M1911 (SAO)

The Colt Military Model has been in service for over one hundred years. This firearm served the US military up until 1985, and even though it was replaced by the Beretta M9, the 1911 still saw service with Special Operations for many years. It's as iconic as any other firearm in this guide.

The 1911 is the first automatic to incorporate all the features we associate with a modern pistol.

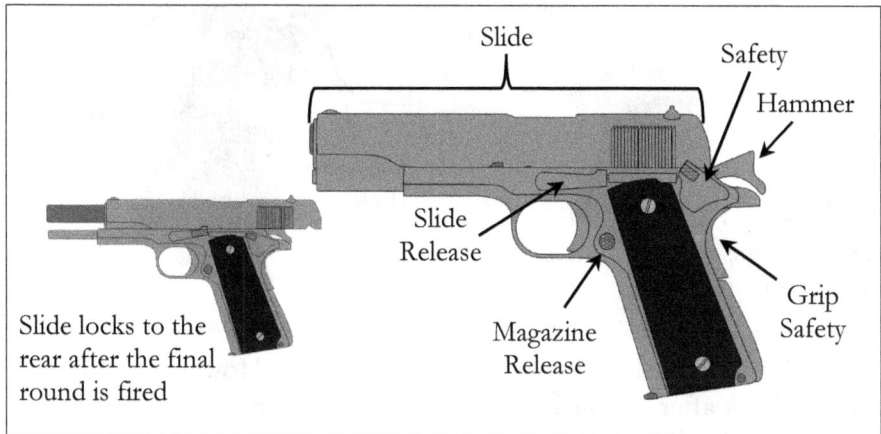

Slide

Safety

Hammer

Slide Release

Slide locks to the rear after the final round is fired

Magazine Release

Grip Safety

The grip safety is an interesting feature. This mechanism prevents the trigger from being pulled if not depressed. This is an effort to prevent an accidental discharge if the gun wasn't being held. Various guns employ this feature, and the grip safety can be either in front or behind the grip.

To Load (Two Methods):
Method One
- Pull back on the slide while pushing up on the slide release – this locks the slide in place for inspection of the chamber.
- Insert a magazine.
- Release the slide.

Method Two
- Insert a magazine.
- Rack the slide to the rear and release.

The slide locks to the rear when the last round is fired.

Walther PP, 1929 (DA/SA)

The PP series by Walther is the second most successful pistol design. This design has been in constant production since it was first released in 1929, and it was the first firearm to incorporate the double-/single-action trigger.

Hammer

De-cocking Safety

Magazine Release

Slide locks to the rear after the final round is fired

The PP series pistols do not have a slide release. However, the slide will lock back after the final round is fired. To reload, insert a new magazine and pull back slightly on the slide. This will release the slide stop and allow the slide to go forward.

Because this is a double-action to single-action trigger, the hammer is back at the end of the firing cycle. Walther needed to invent a way for the shooter to safely de-cock the hammer without pulling the trigger. In this firearm, the safety mechanically de-cocks the trigger when engaged.

To Load:
- Insert a magazine.
- Rack the slide to the rear and release.

Fun Fact: PP stands for Polizeipistole, or police pistol in German.

You can blame Walther for what happens next.

DA/SA Variants

The theory goes that a pistol with a double-action first shot, and the safety disengaged, is inherently safer and faster to get into action than a cocked single-action pistol with the safety on.

A single-action pistol requires a short, light trigger pull. So, to avoid accidents, it must remain on safe until ready to fire, even when holstered.

Because the double-action first shot requires a longer, stronger trigger pull, it can remain off safe when in the holster. Subsequent shots are single action for rapid engagements, and the pistol can be returned to the double-action mode after firing.

The double-action school of thought made sense and inspired a veritable cornucopia of design possibilities.

Welcome to Crazy Town!

Your DA/SA semi-automatic pistol can come in one of three possible variations:

1. De-cocking safety
2. De-cocker/Separate safety
3. De-cocker/No safety

Let's look at an example of each.

De-Cocking Safety (Beretta 92f)

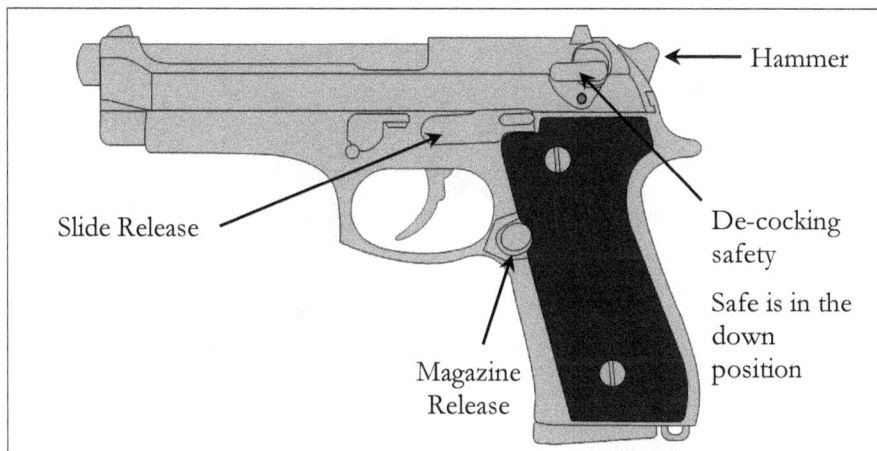

Like the Walther PP, the Beretta has a de-cocking safety. To de-cock, all the shooter has to do is engage, then disengage the safety.

The gun can be carried with the safety on or off, based on personal preference.

De-Cocking Lever and Safety (H&K P30 Variant 3)

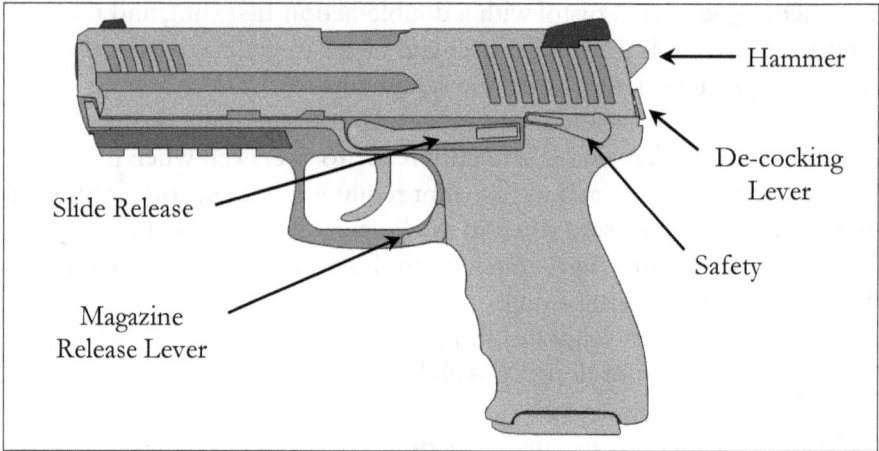

Hammer

De-cocking
Lever

Slide Release

Safety

Magazine
Release Lever

The P30 Variant 3 gives the shooter the same options as the Beretta. However, it removes the need to cycle the safety to de-cock it.

The separate safety allows the gun to be carried in any of the following conditions:
- Single-action with safety on
- Double-action with safety off
- Double-action with safety on

De-Cocking Lever and No Safety (SIG Sauer P226)

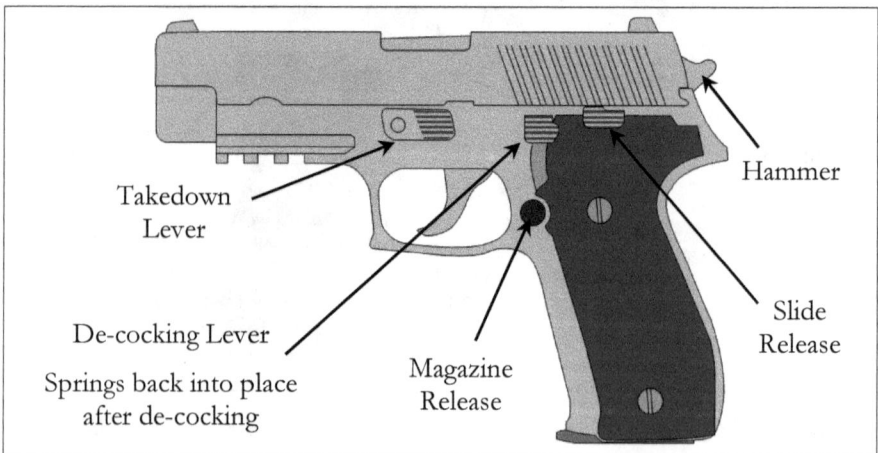

Hammer

Takedown
Lever

Slide
Release

De-cocking Lever

Springs back into place
after de-cocking

Magazine
Release

The SIG P226 has no safety. The safe carrying configuration is with the hammer de-cocked. De-cocking is performed by pressing the lever downward. It springs back into place after lowering the hammer.

Striker-Fired Pistols

Glock isn't the first company to make a striker-fired pistol. Nor is it the first to use a polymer frame. But it is the first, as far as I can tell, to produce one without a manipulatable safety.

Striker is a nomenclature change from firing pin. Striker-fired pistols don't use a hammer to impact the firing pin. Instead, the firing pin is under spring tension and held in place by a sear. When the trigger is pulled, the sear releases and the firing pin strikes the primer, firing the round in the chamber.

Glock 19 (Striker Fired)

Slide Stop

Trigger Safety

Magazine
Release

Glock uses what is called a safe-action trigger. There are several safety mechanisms inside the slide that are disengaged when the trigger is pulled. One selling point of the Glock is that the trigger pull is the same for every shot.

Trigger Safety

Nothing has messed up more writers than the trigger safety. The writer sees a safety listed in the description and they write:

He put the Glock on safe.

Then all hell breaks loose, and there are fifty bad reviews.

In the firearms community, this safety doesn't count as a safety. I know, it makes no sense. It's called a safety, but because it doesn't have visible fire and safe positions, it isn't considered a manipulatable safety.

More and more striker-fired handguns are equipped with this feature. So, just because you see a safety listed, it doesn't necessarily count as a safety.

Machine Pistols

The distinction between a submachine gun and a machine pistol may be hard to define. However, there are some factors that help with the delineation.

First, there's the manufacturer's designation.

Škorpion VZ 61, 1964

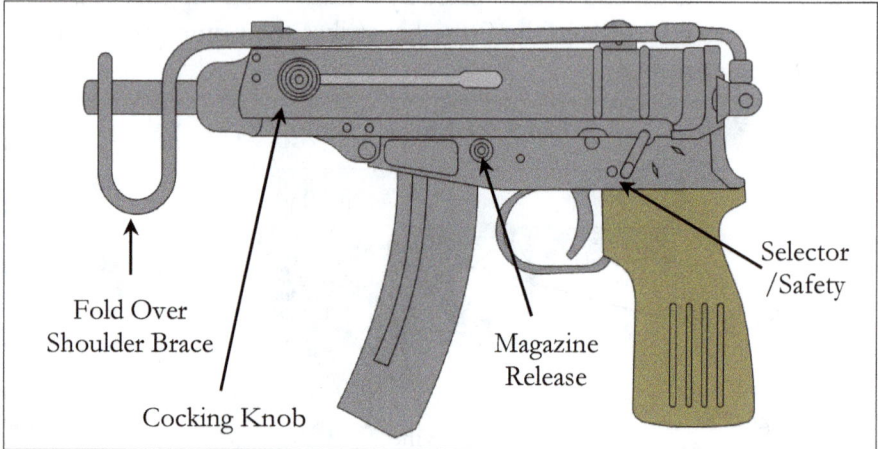

Fold Over Shoulder Brace

Cocking Knob

Magazine Release

Selector /Safety

The Škorpion VZ 61 is designed to be carried in a pistol-like holster and is primarily designated as a machine pistol. Like most of its kind, the Škorpion has an integral stock, or arm brace.

Then we have design.

Glock 18, 1986

Selector

Magazine Release

The Glock 18 is obviously a pistol. It's capable of switching between semi- and fully automatic fire through the switch on the left side of the gun – still no manipulatable safety. The Glock 18 can be equipped with a shoulder stock.

Then there are those that straddle the fence based on how the gun is fitted out.

Ingram M10, 1970

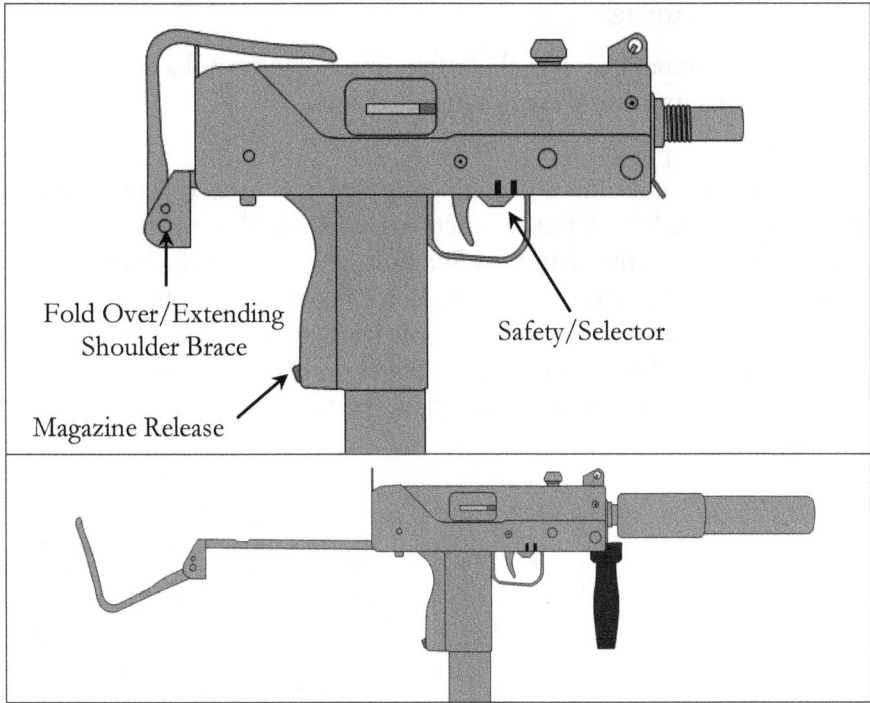

Fold Over/Extending
Shoulder Brace

Safety/Selector

Magazine Release

The Ingram has a threaded barrel to equip either a barrel extension or suppressor. It's more commonly known as a MAC-10, which is chambered in .45 ACP, or a MAC 11, which is chambered in 9x19mm.

I've not fired the Škorpion or the Ingram, but I have fired the Glock 18. If the other two behave like the Glock, then use of the shoulder stock to control recoil is highly recommended.

The design of the Škorpion would lead one to use the magazine as a foregrip. The Glock, with its underside, forward rail mount gives it a foregrip option. And the Ingram has designed foregrip attachments, as many guns of this type do.

Do your research! These firearms are designed for fully automatic fire. This would lead some to assume they're all designed to fire from the open bolt to reduce heat buildup, the Ingram fires from the open bolt, but the Škorpion does not.

The Glock 18 uses a traditional pistol design, as do some others. They fire from the slide forward position. No risk of confusing the way the way they operate.

A word of warning!

Characters unfamiliar with these firearms risk injury if barrel extensions or suppressors are not installed.

Story Time

I used to know a guy who was competent and experienced with firearms. However, he didn't have much experience with machine pistols.

One day, he had the chance to fire an Ingram equipped with a foregrip. He did well firing it at first, then a finger on his support hand crept in front of the barrel. It was simple inattention, but the results were catastrophic. He blew off the pointer finger on his left hand.

Firearms must always be respected, no matter your level of experience.

Timeline of Pistol Development

Deringer – 1825

Collier Flintlock Revolver – c1800

Allen and Thurber Pepperbox – 1837

Navy Colt – 1851, Representative of cap and ball revolver

Lefaucheux M1854 – First cartridge revolver

Sharps Derringer – 1859

Tranter 1863 – One of the first DA revolvers (modern cartridge)

Remington Derringer – 1866

Webley Mark 1, 1887 – The first break-action revolver

Colt – 1889, Creation of the swing out cylinder. All modern revolvers use this design.

Mauser C96 – 1896, First widely used semi-auto – evolutionary dead end

Luger P08 – 1898, First semi-auto pistol to use ejecting mag – evolutionary dead end

M1911 – 1911, Longest lived semi-automatic pistol – still in production

Walther PP – 1929, First DA/SA pistol – still in production

Fun Fact: Raising the pistol to cock – cap and ball shooters would raise their revolvers to the vertical position before cocking to allow fragments from expended caps to fall free and reduce the possibility that they would jam the cylinder. This practice is still seen in media fiction.

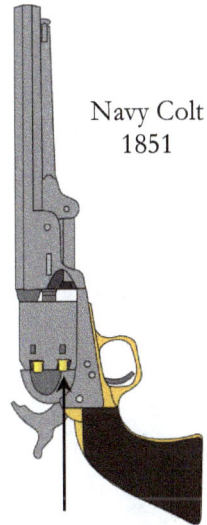

Navy Colt
1851

Percussion caps
would often break
when fired

Operating the Handgun

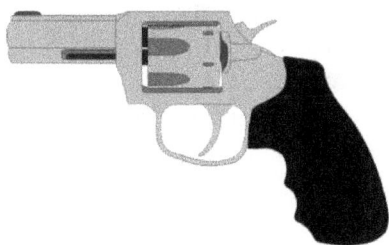

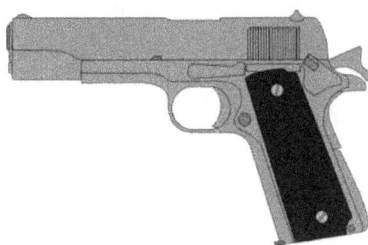

So, how does your character carry their pistol? Well, if you stick with generic actions, it doesn't matter. However, if you want to get down to the nitty-gritty and impress your readers, you need to know the factors that go into deciding what carrying condition your character uses.

Before we get into the conditions, let's look at the factors. The first is easy: training and comfort level. The more experience a person has with a gun, the more comfortable they are carrying it. The more proficiently they can load, unload, and engage targets, the more likely they are to carry the gun and keep it in a condition where it's easy to fire.

The second factor may or may not apply, and that's the agency they work for or were trained by. Agencies have standard operating procedures (SOPs) and policies that govern firearm use. If your character works for a police department, they will definitely follow those policies. However, policies differ between different agencies.

If your character works for a military, the policies and SOPs they follow may differ from unit to unit. Before we get into that mess, let's look at how the revolvers are carried, then we'll tackle the monster.

Carry Conditions, Revolver
Single-Action Revolvers
Single-action revolvers are carried in either half cock with a full cylinder, or hammer down on an empty chamber.

Double-Action Revolvers
As with the single-action revolver, the double-action can be carried hammer down or hammer down on an empty chamber.

Revolvers with safeties – archaic – may be carried with the safety engaged or disengaged, based on shooter preference or agency policies.

Semi-Automatic Handguns
Back in 1976, a retired Marine Corps colonel, named Jeff Cooper, established a weapons training school in Arizona. To simplify talking about how weapons should be carried, he devised three conditions to describe the weapon's status. Since then, two other conditions have been added. If you're researching an agency and their handgun policies, they may refer to the Cooper Conditions.

The internet is also full of firearm aficionados who like to use the Cooper Conditions and advocate a specific condition for a certain type of pistol. This is done to promote their training techniques and their companies. I'll not cast dispersions upon them. The firearms training world is a brutal place, and everyone is looking for their own special angle.

Now, some of these conditions make no sense when applied to certain pistols. In fact, some are downright stupid, depending on the firearm and the proposed condition. So, in the interest of sanity … my own, I'll list the conditions and how they apply to weapons of a certain type.

Theoretically, the lower the number, the fewer steps you need to pull the trigger. However, this isn't always the case. As we will see. Consider all magazines as full.

Possible Carry Conditions

	Chamber	Magazine	Hammer	Safety	Applies to
4	Empty	Not Inserted	Down	N/A	All
3	Empty	Inserted	Down	N/A	All
2	Loaded	Inserted	Down	On/Off *	DA/SA
1	Loaded	Inserted	Cocked	On	SA
0	Loaded	Inserted	Cocked	Off	All

Condition 4: This is an empty gun. The magazine is out, and the chamber is empty. Simple enough, every gun can be in this condition.

Condition 3: The magazine is inserted, but the chamber is empty. This condition is applicable to all semi-autos. This Condition is also known as Israeli Carry.

Quick Aside: When Israel first became a country, it had a whole bunch of pistols of varying quality and design from all over the world. To standardize training, all pistols were to be carried with a loaded magazine, chamber empty. This practice holds to the time of this writing. However, some units now operate with fully loaded handguns.

Condition 3 is for those who are uncomfortable with their firearm or want an extra step before the gun can be fired.

Condition 3 is also known as cruiser-ready. This is a term used to describe the status of a shotgun or rifle. (Page 113)

Condition 2: The magazine is in, a round is chambered, and the hammer is down. This condition should only apply to DA/SA pistols. It's how these guns were designed to be carried.

It's possible to put a single-action pistol in this condition with the safety off. I've seen some people argue in favor of it. However, it makes no sense. Clicking off the safety is always going to be easier and faster than cocking the hammer. Additionally, single-action pistols aren't equipped with a de-cocking mechanism. This means the shooter must manually de-cock the gun themselves, a situation ripe for accidental discharge.

*In DA/SA pistols, the safety can be on or off, depending on agency SOP or personal preference.

Condition 1: The magazine is in, the chamber is loaded, the hammer is back, and the safety is engaged. This condition applies to all single-action semi-autos. It's how the firearm was designed to be carried.

Certain DA/SA pistols are capable of being in this condition. Whether or not they're carried that way will be a matter of personal preference and agency SOP. See H&K P30 Variant 3. (Page 166)

Condition 0: The magazine is in, the chamber is loaded, and the hammer is back. This condition describes any gun you're about to shoot.

One could argue that any double-action-only, or striker-fired pistol is in this condition as soon as it's loaded.

Okay, time to stop the silliness. First of all, unless your character is a range master or firearms aficionado, they aren't going to use this terminology. Second, even when I performed the duties of a range master, I never instructed the firing line to "Go to condition four."

If I wanted everyone to clear their weapons I'd say, "Clear your weapons and buddy check." You can't take for granted that everyone knows what Condition 4 is.

Scenario: A person is talking about how they carry their pistol.

"I carry my Glock in Condition 3. It's safer this way."

Dumb. Who writes these examples? Oh yeah, me. Do you really want your reader to stop and go look up what Condition 3 is?

"I carry my Glock Israeli style. It's safer this way."

No, for the same reason as above. Unless you're about to, or already have, defined Israeli carry.

"I carry my Glock with a mag in, but the chamber clear, It's safer this way."

Success! I finally got it right.

Caveat: Using the terms Condition 3 and Israeli carry may fit the character. Be sure to explain the terminology to the uninformed reader.

Another note on Condition 3: a police agency may require their officers to carry their pistols Israeli style, although I haven't seen it here in the US. Similarly, a military installation or base could require their police force to carry their handguns in Condition 3. This I have seen on a US base overseas.

IF you decide to use Condition 3 as a plot point for a soldier or police officer, then spend a few words on why. This will help reduce negative review comments. Example:

After the shooting incident last year, the new post policy required MPs to carry their pistols Israeli style. She hated having an empty chamber …

You don't need any further detail, it's a post policy, done.

Carry Condition Summary

The following table lists how the firearm should be carried if not in Condition 3.

Firearm	Configuration
Single action Revolver	Loaded / Half cock
Double action revolver	Loaded / Hammer down
Single action Semi-automatic pistol	Loaded / Hammer back/on safe
DA/SA semi-automatic pistol	Loaded / Hammer down/Safety on or off. (Contingent on agency SOP or preference)
Striker fired and DAO guns	Loaded

I could spend a little time ranting here about how ridiculous the Conditions are, but I won't. The bottom line is that the gun's manufacturer designed it to be carried a certain way. If you carry it differently, you do so at your own peril.

The real issue, for writers, is the status of the safety:
- Is the gun equipped with a safety?
- Is the safety supposed to be engaged by design? For single-action pistols, yes.
- Is the safety supposed to be engaged per agency policy? Depends.

I've got a perfect scenario to illustrate how use of the safety can be a trap for writers. But before I do that, we have to talk about …

Holsters

All handguns should be carried in a holster. Anyone who carries a handgun in their pocket or tucked into their belt is an idiot.

The holster serves two purposes: it makes the gun secure and protects the trigger.

I'll allow carry bags and purses if there is *nothing else* in the pocket that hosts the gun. If your character keeps their pistol in a purse or shoulder bag, just rattling around with keys, makeup, and other personal detritus, bad things can happen. Your character may fumble the draw, or there could be an accidental discharge. On the other hand, you may want this to happen.

Having said all that, have I been an idiot? Yes, on a few occasions. Sometimes you don't have the proper holster for a given situation. A tactical leg holster is the wrong look for a formal event. There's no one holster that's perfect for every circumstance, and if your character doesn't have time to get the right holster, they may need to improvise.

Holsters can be as simple as a slide holster, which is a length of firm material formed into a loop to hold the gun and protect the trigger. These can be attached to a belt or bag by various means.

Holsters can also be complex, with varying levels of retention that must be manipulated by the operator before the pistol may be drawn.

Retention is a good thing; it keeps the gun from clattering on the floor when your character bends over.

These holsters include shoulder, flap, concealed carry, and tactical. The available options almost outnumber the possible variations of handguns. Each option has its own pros and cons.

Bottom line: If your character is military, law enforcement, or experienced, they'll use a holster if at all possible.

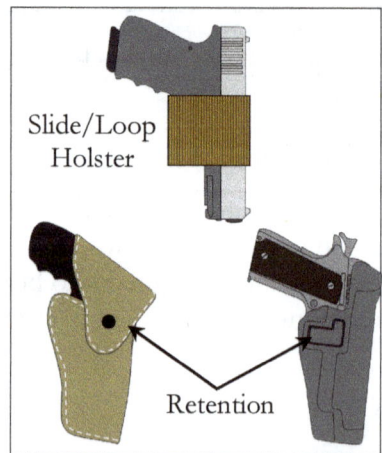

Slide/Loop Holster

Retention

Inexperienced or untrained characters may just stick the pistol in their belt or a pocket.

Concealed Carry

If your character is carrying a concealed firearm, they should think about printing. Printing is an informal but widely used term that means the location of carry, weapon size, or holster results in a noticeable bulge or outline.

The detective noticed the bulge at the small of the suspect's back. "He's armed."

Printing usually occurs with light or tight clothing. Looser, bulkier clothing reduces the effect. Of course, all concealed weapons will print at some point. They tend to reveal their outline when someone bends over, crouches, or sits.

The Safety and the Holster

So, we have all these different types of guns with varying trigger and safety configurations. Your next question should be, "Well, if the gun is in a holster, and the trigger is protected, why engage the safety?"

And you have a good point because pistols like the Glock and the SIG P226 don't have safeties. It all comes down to training, safety, and SOPs.

There are pros and cons for every weapon, and some are considered *safer* than others from a liability perspective. Likewise, use of the safety is measured from a liability standpoint.

Let's look at the Beretta 92 versus the US Army.

Note: The Beretta 92 was the US Army's service pistol from 1985 to 2017. It's the one I carried.

Special Operations soldiers receive a lot of training and fire their weapons more often than regular soldiers. Their carrying configuration in the holster is:

• Loaded, hammer down (double action), safety *OFF*.

Regular soldiers don't often carry a pistol. With less training than the Special Operations guys, their configuration in a combat zone is:

• Loaded, hammer down (double action), safety *ON*.

This is the trap. Readers, especially those with training, will default to that training. They may not recognize or appreciate the policies of another organization.

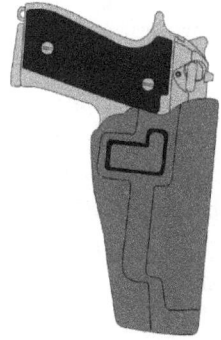

Story Time

I was in Afghanistan doing laundry. I was armed and my Beretta was in a slide holster on my belt. A sergeant from another unit noticed my pistol was on fire. He decided to point this out and give me a lecture on firearms safety. I politely informed him that I was following *my* unit's SOP and if he had a problem with that, he could talk to my battalion sergeant major. I turned around and continued folding my socks.

I never heard anything else about the incident. No doubt he's out there thinking that Special Forces guys are a bunch of cowboys.

Yippee ki-yay ...

The moral of the story is this: that sergeant may read fiction and *know* the proper way to carry a pistol is the way he did in the Army.

Moving While Armed

You will probably never write this. In fact, never even mention this in an action scene. This is one of those trivia-type situations, information only for your awareness.

If your character is armed with a pistol in a running gun battle, or possibly a chase, they should make the gun as safe as possible before moving. Single-action revolvers should be set to half cock. Single-action automatics should be put on safe. And DA/SA guns should be de-cocked. Rifles and shotguns should be put on safe too.

When moving, running, and dodging, it's easy for the trigger finger to slide into the trigger guard ... then bad things happen. Your characters should make the gun as safe as possible, yet still be ready to fire in these action scenes.

This calls back to the most recent scenario. If the scene is to demonstrate competence, you may want to write in these detailed actions. Otherwise, leave them out and let the reader assume the character is moving safely.

Emergency Reload

An emergency reload occurs when the shooter has fired the gun until it's empty and now must reload it as quickly as possible. A tactical reload, as mentioned on page 127, is when ammunition is added, or a magazine is replaced prior to the firearm going empty.

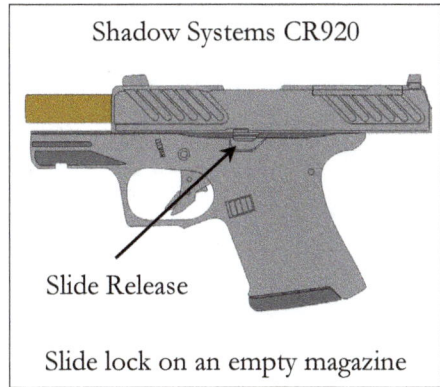

Shadow Systems CR920

Slide Release

Slide lock on an empty magazine

It's the emergency reload that we will focus on here – specifically, for semi-auto pistols. As you know, with most semi-auto pistols, the slide locks back on an empty magazine. After the shooter has replaced the empty magazine with a full one, they have two options to release the slide.

The first option is the **slide release method**. The shooter simply depresses the slide release mechanism, and the slide will move forward and chamber the first round in the magazine.

Option two is called the **slingshot method**. In this technique, the shooter grasps the slide with their non-firing hand, pulls back on the slide – which releases the slide stop, and releases the slide. The slide then moves forward and chambers the first round in the magazine.

The slingshot method derives from pistols without a slide release mechanism, like the Walther PP.

Which technique is best? Well, that's a hotly debated subject in all the gun forums. Do an internet search and you'll find thousands of results extolling the virtues of one, the other, or both. I have my own opinions, and if we meet at a conference, I'll gladly share them with you. However, this is not the place for that discussion.

Note: In performing the slingshot technique on certain pistols, there is a chance the shooter may accidentally engage the safety mechanism.

Ammunition

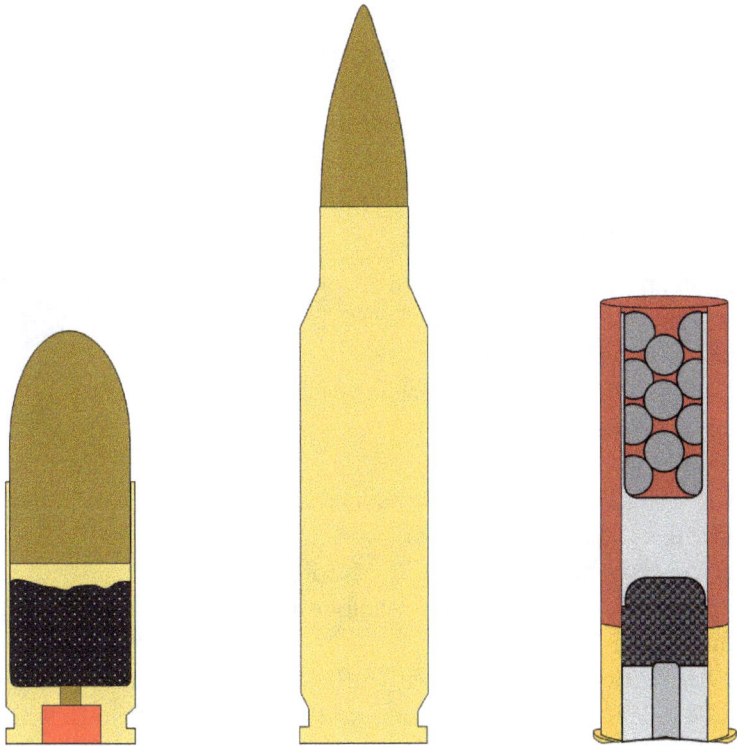

Bottom line up front: You don't need to know everything there is to know about ammunition. Is it good to know? Sure, but much like a minor character's back story, most of this information belongs in your references file. Any research you do for a firearm will yield the caliber for that gun.

The type of ammunition the gun can use is usually stamped on the gun, and the caliber information is stamped on the ammunition's casing.

Most writers can, and should, stop here.

However, for some, ammunition confusion can be an important plot point. Recently, I had an interesting gun/caliber problem to solve. The parameters were:

- A pistol that Ernest Hemingway could have owned.
- Confusion in the type of ammunition it used.

FN Model 1910

The gun and its ammunition are an important plot point for this murder mystery. It was a fun problem to solve. Eventually, I found the perfect candidate. I present to you the gun that Gavrilo Princip used to assassinate Archduke Ferdinand and kick off WW1, the FN Model 1910, also known as the Browning 1903.

It can be chambered in .32 ACP or .380 ACP.

The funny thing about the .380 caliber, this round is also known as 9mm Short.

The victim, being a Hemingway buff, refers to the ammo as 9mm Short. The killer buys *standard* 9mm ammo. But standard 9mm will not fit inside a gun chambered for .380 ACP. Plot point achieved.

The following pages will attempt to unravel some of the confusion around the various types and sizes of ammunition. And there's a lot of it.

Naming Conventions

As you may recall, the problem starts all the way back to when breech-loading firearms were first developed. There were no standards. Hell, there wasn't even ammunition. If an inventor wanted to create an automatic pistol, they first had to create the cartridge it would fire.

Let's start with what should be the easy stuff. Let's say you want to design a cartridge. The bullet needs to fit in the bore. If you remember, the bore has rifling made up of lands and grooves. Which measurement are you going to use, the distance between the lands or the distance between the grooves? It could be either, depending on who manufactures the ammunition.

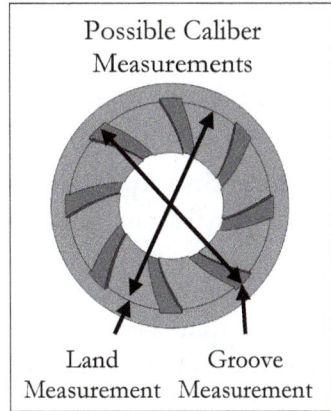

Possible Caliber Measurements

Land Measurement   Groove Measurement

Whichever measurement you use, you now have the caliber. This can be expressed in millimeters (metric) or in hundredths of an inch (imperial).

The bullet fits in the cartridge and the cartridge needs to fit in the chamber, so now you need the size of the casing. This is called headspace, and it's a measurement from the face of the bolt to some feature of the casing. In straight-walled ammunition, like pistol, it's usually the length of the casing. With bottle-necked ammunition, like most rifle rounds, the measurement can be taken at the shoulder, neck, or rim.

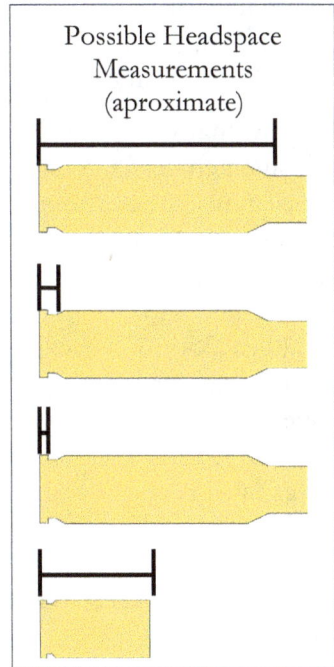

Possible Headspace Measurements (aproximate)

Ammunition using the metric system is great at defining the characteristics of a round of ammunition. The first number is the caliber and the second is the size of the casing: 9x19, nine-by-nineteen. 7.62x51, seven-six-two by fifty-one.

Ammunition using the imperial measurement system sucks. The second measurement is a cumbersome beast when it is used. I'll use ammunition for an AR variant as an example: .223x1.4640.

So, instead of using such a terrible expression, most ammunition (imperial) uses an appellation to designate the size of the casing. This is usually based on the original manufacturer of the round: .223 Remington.

Some of these appellations make sense. For instance, ACP stands for automatic Colt pistol.

But others don't. When you look at .30-06 (pronounced: thirty-ought-six), you would think that the "ought-six" part refers to some aspect of the bullet or cartridge. You'd be wrong. Springfield designed its .30-caliber rifle cartridge in 1906. Hence, the name. Fail in the naming category, if you ask me.

Ammunition Appellations

For imperial ammunition, the appellation is critical for choosing ammunition that will fit in the gun. However, metric ammunition uses appellations too. 9x19 ammunition often has Parabellum, Para, or Luger attached to it. This can be an important distinction because 9mm Makarov and 9mm IMI will not fit in guns designed for 9mm Parabellum/Para/Luger.

When firearm aficionados talk about common ammunition, the appellations are often dropped. Thus:

- 9x19mm Parabellum/Para/Luger becomes 9mm
- .30-06 Springfield/SPRG becomes *thirty-ought-six*

Now that you know about the headspace measurement and the importance of appellations, we can reexamine the ammunition confusion between the .380 ACP (9mm Short) and 9mm Parabellum in the example that started this chapter.

When .380 ACP is expressed in millimeters, it's 9x17mm, or 9mm Short. There are other appellations for the short round.

9x19mm is a much longer cartridge and will never fit in a firearm chambered for the smaller 9x17mm round.

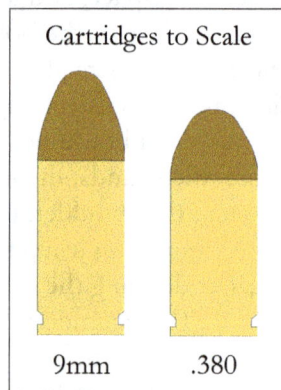

Cartridges to Scale

9mm   .380

The same goes for rifle ammunition.

7.62x51mm NATO is not compatible with the Russian caliber 7.62x39mm.

Because headspace information and appellations are so often dropped in conversation, firearm novices tend to think that guns with the same caliber fire the same ammunition.

Other appellations are important for safety. You may encounter ammunition with the following:

+P, +P+, Express. Ammunition with these designations is loaded to achieve higher pressures and deliver greater velocity to the projectile. Firing these rounds through a gun that is not designed to handle those higher pressures can result in damage to the gun and injury to the shooter. There is no industry standard guiding these designations.

Magnum is another appellation with ill-defined standards. A magnum round is also loaded to achieve higher pressures and sometimes has a heavier projectile. The saving grace of the magnum designation is that it also defines a larger cartridge size. .22 Magnum ammunition will not fit in a gun designed for .22 LR.

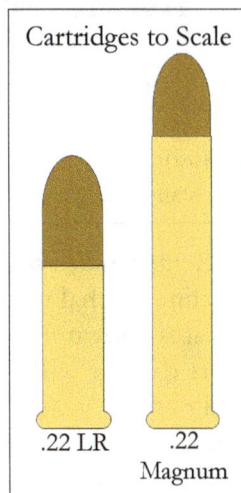

Cartridges to Scale

.22 LR   .22 Magnum

Match Grade is another designation you'll see on an ammo box. This term is also poorly defined because there is no industry standard that guides it. Theoretically, match-grade ammunition is manufactured to a higher standard with tighter tolerances.
(See page 252 for rules on writing caliber.)

Projectiles

Before we get into the characteristics of bullets, let's talk about what they can and can't do. Arrows or crossbow bolts have two tasks: travel a distance and penetrate the target. The resulting penetration caused injury or death.

To perform the same tasks, a bullet needs to go really fast. Velocity is what gives the bullet its ability to travel a distance and penetrate the target. However, that velocity has a downside – bullets tend to overpenetrate the target or, in other words, go through it. Some bullets are predisposed to overpenetrate. Others are designed to reduce that possibility.

When the bullet enters the body, it creates a permanent cavity by punching through the flesh. It also creates a temporary cavity as the result of hydraulic pressure.

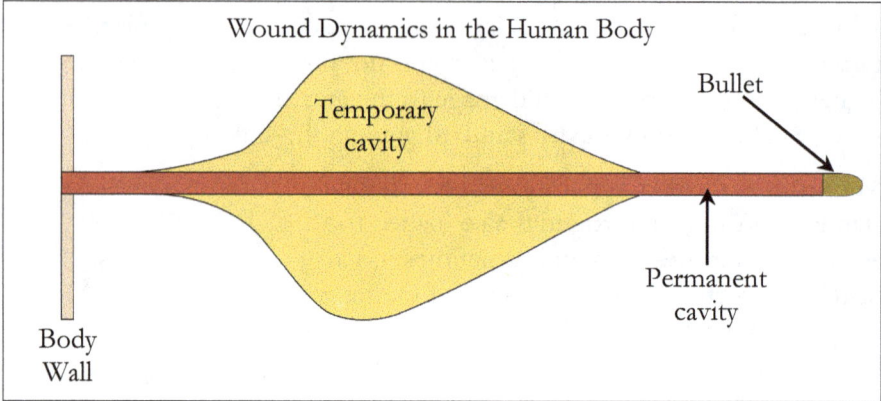

Wound Dynamics in the Human Body

Some studies suggest that the temporary cavity is not as integral to creating a lethal wound as the permanent cavity caused by the bullet.[4] What is known for a fact is that bullets that expand and fragment cause more damage when they do.

I'm not a wound forensics expert, but what I can say is that if a bullet starts to tumble or expand on its way through a body and it then exits the body, the hole going out will be bigger than the one where it went in. Bullets also break bones and sever tendons.

Stopping Power

Caliber snobs believe that bigger, heavier bullets moving at high velocity are the biggest factors in stopping power. They tend to discount any ammunition smaller than .45-caliber and associate stopping power with lethality.

Stopping power is the ability of the weapon to incapacitate or immobilize the target. Shooting a person in the leg with a .45 doesn't have the same stopping power as cracking open their skull with a lead pipe. But, of course, I can shoot a person from a safer distance.

So, stopping power doesn't equal lethality, but lethality does equal stopping power. And any caliber can be deadly. There's a meme floating around out there that the .22-caliber cartridge has killed more people than any other. I don't know about that. What I can tell you is that inside 100 yards, the .22 can be a deadly round.

---

[4] "Wound Ballistics 2-Mechanics of Projectile Wounding," Viper Blog, Viper Weapons, August 18, 2023, https://viperweapons.us/blog/f/mechanics-of-projectile-wounding

You want to know what kills people more than caliber or bullet type? Shot placement. If you shoot a person in a vital spot, they will die. It doesn't matter what kind of ammunition you're using.

Wyatt Earp said it best, "Fast is fine, but accuracy is final."

This brings us to capacity. Guns with larger calibers tend to have a smaller capacity than guns with smaller calibers. The 1911, .45-caliber, has a standard seven to eight round-magazine. The Beretta 92f, 9mm, has a standard fifteen-round magazine.

I, personally, would rather have fifteen opportunities to shoot someone with my Beretta 92, than a paltry seven chances with a 1911. In my mind, capacity trumps caliber.

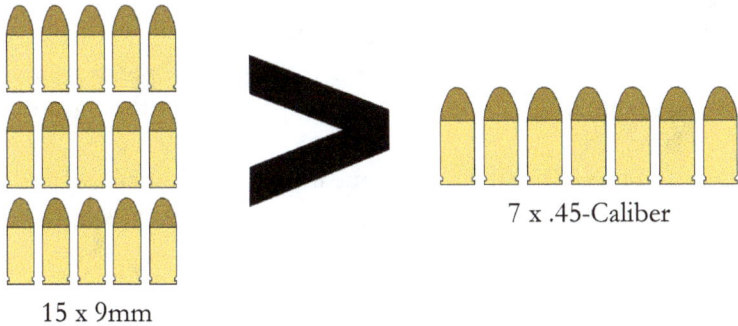

7 x .45-Caliber

15 x 9mm

Sorry, got off track for a minute, but it all melds together sometimes.

What bullets can do: They penetrate (and sometimes overpenetrate) the target, either wounding or killing them.

What bullets can't do: pick up a person and throw them across a room.

Scenario: The character is armed with a sawed-off shotgun loaded with 12-gauge slugs. He is engaging his nemesis.

He fired from a distance of five feet. The slug punched into the villain's chest throwing him backward into the wall.

Not even remotely correct. There is no way any projectile is going to hurl a body any distance at all. First, we already know the slug is going to penetrate. This means that not all the kinetic energy of the projectile is passed on to the target.

Next, let's talk about physics. Newton's third law states that, for every action, there is an equal but opposite reaction. This means that if the force of the projectile was great enough to throw the villain against the wall, the hero would have been thrown backward with the same force. No matter how hard that gun kicks, it won't throw you across the room.

Okay, a tiny person *can* get knocked back by a 12-gauge shotgun, but not like you see in the movies.

Here's the good news. You are the god of your universe. Apart from violating the laws of physics, you can make a shot as lethal as you want.

It is known that people who suffer mortal wounds can continue to act until they succumb from them.

Do a quick search for the Miami Shootout, 1986, and read the account. Of the ten people who participated in the gunfight, only one emerged unscathed. The fight lasted five minutes and more than 145 rounds were fired.

One of the bank robbers, Platt, received a shot through his right lung early in the shootout. This was a mortal wound, but he fought on until his spinal cord was severed at the end of the fight.

A determined person can fight through the pain of being shot. They can find ways to manipulate and reload weapons when their arms and hands are damaged.

I've said it before, but it's worth repeating here: shot placement matters more than caliber or bullet type when it comes to stopping your foe.

Pistol Bullet Design and Construction

A bullet is generally constructed of a lead core encased in a copper jacket. The jacket protects the barrel from deposits and fouling. The jacket also prevents the bullet from deforming in the barrel.

Ball ammunition uses bullets with a full metal jacket (FMJ). This term is a nod back to the original projectiles. Ball ammunition tends to overpenetrate the target because the full jacket prevents the projectile from expanding.

This bullet is going to go through the target and keep on trucking. For this reason, ball ammunition is usually used for target practice and training on the firing range.

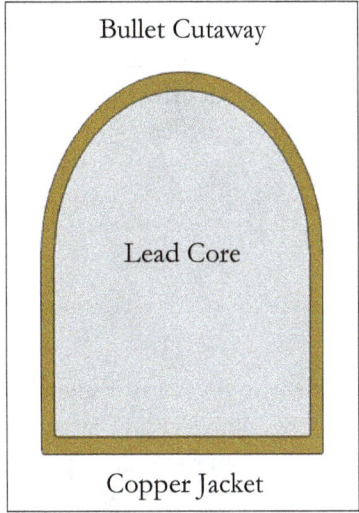

Bullet Cutaway

Lead Core

Copper Jacket

Semi-jacketed bullets are an attempt to solve the over-penetration issue.

The soft lead core is exposed allowing for good expansion while keeping part of the jacket to reduce fouling.

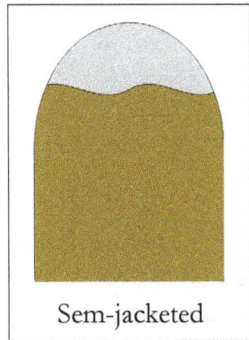

Sem-jacketed

Hollow point bullets have a concave nose and are designed for maximum expansion when they hit a target.

Hollow points come in both fully jacketed and semi-jacketed varieties.

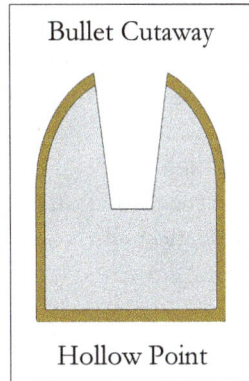

Bullet Cutaway

Hollow Point

191

Wadcutters and semi-wadcutters are good bullets for target shooting. These have flat noses designed to cut cleanly through paper.

Wadcutters are generally made of lead. Semi-wadcutters can be fully jacketed, semi-jacketed, or lead.

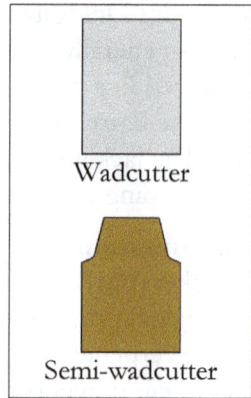

Wadcutter

Semi-wadcutter

Armor-piercing bullets have a hardened metal core, such as steel or tungsten. They're encased in a copper jacket to prevent damage to the rifling in the barrel.

Armor-piercing rounds are available in handgun calibers, but they're illegal in the US.

Bullet Cutaway

Armor Piercing

Tracer ammunition is used to observe fire, direct fire, or as a round count indication tool.

The tracer sacrifices part of its core to make room for a pyrotechnic material.

When the round is fired, the powder charge ignites the pyrotechnic, making the trajectory of the bullet visible, even in daylight.

In machine gun ammunition, the distribution of tracers to regular ammunition is 1:4, one tracer for every four regular rounds.

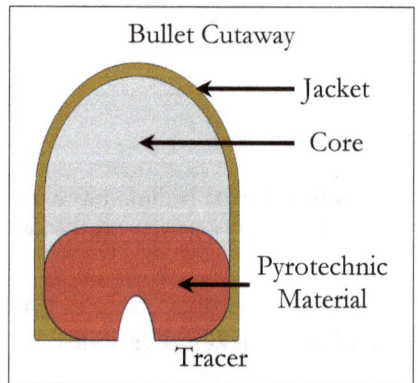

Bullet Cutaway

Jacket

Core

Pyrotechnic Material

Tracer

As a tactical tool, soldiers and operators sometimes reserve a magazine full of tracer rounds. In a firefight, the leader can use this magazine to direct the fire of other team members or mark a target for air support.

When used as a round count indication tool, operators will load tracers in a 4:1 ratio with the last five rounds as tracer. This lets them know where they are in their magazine.

Funny thing about tracers. They work both ways. Your friends can see where the rounds are going – and the bad guys can see where they came from.

Frangible bullets are made from a compressed powdered metal like copper but can also be made of zinc or tin. The bullet disintegrates when it strikes a hard target like steel.

These rounds are used for competitions where shooters engage steel targets.

Special Ammunition

There are a ton of options on the market that are variations of the basic designs listed above. All have their pros and cons. The most interesting handgun load I've found is the Glaser Safety Slug, which is designed to reduce the possibility of overpenetration.

The projectile is of hollow point design and has been filled with small pellets. The projectile is capped with a polymer tip. Upon entering the target, the polymer tip causes the bullet to expand, releasing the pellets.

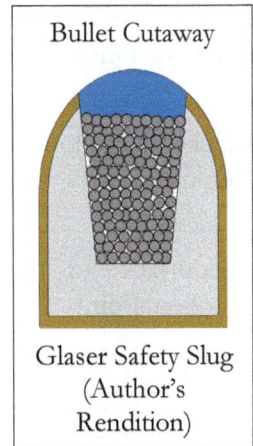

Bullet Cutaway

Glaser Safety Slug
(Author's Rendition)

Blanks

Blanks are cartridges with no projectile. They're used in movies and training to simulate gunfire. Some people think that blanks are harmless, but they're not safe. There may not be a projectile, but hot gas still leaves the barrel under high pressure – and several actors have been killed by blank ammo.

Which Type Should You Use?

That all depends on the nature of your character. If your character is in the military, certain conventions restrict the use of expanding bullets, so no hollow points for you.

On the other hand, if your character is a police officer, they may be armed with hollow points to reduce the possibility of overpenetration and injuring bystanders.

If your character is not a cop or a soldier, the floor is wide open. I've compiled a table of possible bullet types for your convenience. It's not an all-inclusive list, but it does hit the most common types.

Name	Abrv	Characteristics
Full Metal Jacket (Ball)	FMJ	Soft lead core fully encased in copper
Soft Point	SP	Soft lead core partially encased in copper
Hollow Point	HP	Hollow tipped bullet designed to expand on impact
Jacketed Hollow Point	JHP	Hollow tipped bullet designed to expand on impact. Jacket covers the entire exterior of the bullet
Semi-Jacketed Hollow Point	SJHP	Hollow tipped bullet designed to expand on impact. Jacket covers most of the exterior surface of the bullet
Wad Cutter	WC	Flat nosed bullet
Semi-Wad Cutter	SWC	Partially flattened nosed bullet
Armor Piercing	AP	Hardened core bullet designed to penetrate steel
Tracer		Bullet with a pyrotechnic in its base which burns when fired
Frangible	FRNG	Compressed copper bullet designed to disintegrate on impact
Blank		A cartridge with no projectile. Usually sealed with wax

Rifle Bullets

Rifle bullets are longer and thinner than pistol bullets, and they build higher velocities in the longer barrel, which gives them greater range. They also have greater penetration capabilities as a result of their shape.

Except for wadcutters, rifle bullets come in the same types as handgun bullets (refer to previous page). However, there is one variation available in rifle bullets that handgun ammunition can't offer: the boat tail.

Here we have yet another hotly contested subject in the gun forums.

Boat tail projectiles have a tapered base that's designed to reduce air resistance and increase a bullet's ballistic coefficient. This allows them to fly through the air more easily.

Flat-based bullets have a closer resemblance to handgun bullets.

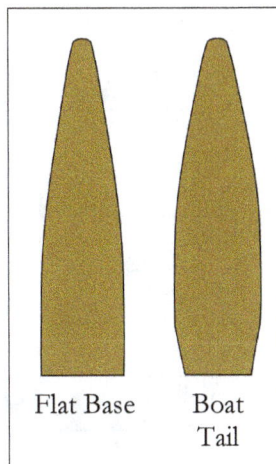

Flat Base   Boat Tail

Studies have been performed to answer the question of which shape is best.[5] They've ranged from super scientific to a bunch of dudes posting about their day at the range.

In addition to a bullet's shape, there are a ridiculous number of factors that affect a bullet in flight. Air temperature, humidity, winds, and atmospheric pressure all combine into a morass of uncontrollable variables. Precision marksmen try to account for these factors when planning a shot.

As noted in the Zeroing Procedure section (page 120), snipers keep a log of almost every shot they take. They review this log to note the conditions of a previous shot and their results. They also note whether the shot was with a cold or warm bore. The results of a cold bore shot (the first shot through the gun) will be slightly different than the results of a warm bore shot (any round fired after the first). Shooting is an art as much as it is a science.

If you want to really nerd out, talk to a sniper about what type of ammunition they prefer.

Okay, we're done with the precision guys for a while. Let's talk about the people who shoot in bulk. These folks aren't as concerned with the finer points of shooting. They're out on the two-way rifle range trading hate with the bad guys. These are soldiers, cops, and their various Special

---

[5] "Scott E. Mayer, "Banging Out Boattails," RifleShooter, January 4, 2011, https://www.rifleshootermag.com/editorial/ammunition_rs_boattails_093009wo/84178.

Operations units. These folks have an array of ammunition types not available to your average civilian, and they all serve a special purpose.

The US military lists ammunition and its purpose by color code. Other countries do the same. However, their color codes are different. You guessed it, no international standard.

Color codes are found in the packaging and on the nose of the round:

Color	Bullet type
Red	Tracer
Black	Armor-piercing
Silver	Armor-piercing/incendiary
Green	Steel core for penetration against hard targets
Blue	Incendiary
Yellow	Observation (flash & smoke on impact)
Yellow-Red	Observation/tracer
Orange	Dark ignition tracer
Purple	Infrared Tracer

Not all these ammunition types are used by police officers. They probably won't use any of the tracer variants because of their potential to start fires. Cops might use green tip or armor piercing, but that will be a policy decision. I'll go out on a limb here and say no, due to the tendency of those rounds to overpenetrate the target. Police agencies avoid this as much as possible … liability.

Your standard ammo for soldiers will be green tip or ball. Cops will use ball or hollow point … most likely ball.

Propellants

Traditional gunpowder is made up of three components. A nitrate, which provides oxygen; charcoal, which provides fuel; and sulfur, which is a fuel that also reduces ignition temperature.

Modern recipes are vastly different. Forgive me if I don't go into detail on this; I'm not a chemist. Suffice it to say, the recipe for firearm propellant has changed quite a bit since it was first discovered.

I will say this about propellants because it affects you, the writer, cordite was an ingredient used in gunpowder between 1889 and 1941, but it's no longer an ingredient for firearm propellants. Care should be taken when referring to the smell in the air after a gunfight. Unless your story occurs in the date range above, don't use "the smell of cordite" in your writing.

Interlude: Body Armor and the Cop Killer Bullet

The age-old battle between the tools humans use to inflict damage on each other, and the measures used to protect against that damage, has raged since antiquity. It started with shields and progressed through helmets and armor.

Strategy, tactics, and battle formations are nothing more than a way to reduce the effects of an enemy's weapons and increase the effectiveness of your own. The firearm changed a lot of that. For a long time, nothing could effectively stop a bullet, so armor became an obsolete burden and was discarded.

With the advent of modern materials, body armor has regained its relevance. As described by the National Institute of Justice (US), ballistic body armor falls into five classes.[6] Each class is designed to offer protection against the listed calibers.

- Level 2A: Low velocity 9mm, .40 S&W
- Level 2: High velocity 9mm, .357 Magnum
- Level 3A: High velocity 9mm, .44 Magnum
- Level 3: Rifle rounds up to 7.62 Lead core
- Level 4: Armor piercing rifle rounds

These designations address a certain caliber at a specific velocity, with a limited number of bullet impacts. The material can't sustain damage forever. To achieve Level 4 status, armor need only protect against *one* impact of armor-piercing rifle ammunition. However, many companies offer products that offer multi-hit protection.

Level 3 body armor consists of a lightweight vest that contains bullet-resistant material. Newer materials have replaced Kevlar. So, be careful in referring to a "Kevlar vest."

To attain Level 3 or Level 4 protection, a plate must be added. These plates are often made of ceramic, but they can be of other materials.

The plate fits into a pocket on the vest, and some tactical rigs can contain everything including attachment points for ammunition carriers. There are many possible configurations.

Body armor doesn't protect all your favorite parts to the same level.

---

[6] Justice Technology Information Center, "Understanding NIJ 0101.06 Armor Protection Levels," US Department of Justice Office of Justice Programs, accessed April 20, 2024, https://www.ojp.gov/pdffiles1/nij/nlectc/250144.pdf.

The figure to the right is only protected to Level 3A in the yellow area. The green area shows where the figure is protected to Level 4. Everything else is unprotected.

Police officers wear 3A vests under their uniform when on duty. They may wear lightweight plates that improve the armor to Level 3, but Level 4 plates are deemed too bulky for casual use.

"Cop killer" rounds are handgun ammunition designed to defeat ballistic armor Level 3A and below. This includes hardened core bullets and/or projectiles with a Teflon coating. This ammunition is illegal in the US, but it does exist.

According to FBI 2021 statistics, 61 officers were killed by firearms:

- 15 by handguns
- 11 by rifles
- 2 by shotgun
- 33 by an undetermined or unreported firearm[7]

Any and every handgun caliber from the lowly .22-caliber to the ridiculous .50-cal is a potential cop killer. Body armor is good, and it saves lives, but nothing stops the right bullet in the right place. Not for long anyway.

Should your antagonist use armor-piercing/cop killer ammunition?

Is your character the type who would risk exposure by carrying armor-piercing handgun ammunition? In some jurisdictions, illegally carrying a concealed handgun is a misdemeanor, but carrying armor-piercing handgun ammunition is a felony in almost every jurisdiction.

Is the use of this type of ammunition important to the plot? If the ammunition is encountered by the protagonist, it had to have come from somewhere. Do you need this link in the chain that leads to your final conflict? Spoiler: this was the plot for *Lethal Weapon 3*.

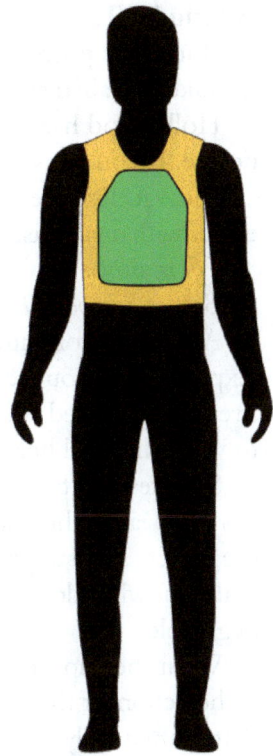

---

[7] "Crime Data: Law Enforcement Officers Killed in the Line of Duty Statistics for 2021," Law Enforcement Bulletin, FBI, November 9, 2022, https://leb.fbi.gov/bulletin-highlights/additional-highlights/crime-data-law-enforcement-officers-killed-in-the-line-of-duty-statistics-for-2021.

Interlude: The Suppressor

Nothing says professional killer like a suppressor. They are often mistakenly called silencers by the novice or inexperienced.

Hollywood has developed its own sound for rounds fired through this device. If you pause for just a moment and quiet your inner voice, you can hear it. I've seen writers try to accomplish this Hollywood sound effect with a surprising array of consonants. The most popular being, *Pfft*.

Pfft is almost always done in italics ... because they're quieter.

Bottom line, oops, I'm still whispering.

Ahem. Bottom line up front, just keep *Pfft*ing to your heart's content. Explaining to your reader what it actually sounds like when an operator fires a suppressed rifle or pistol takes too many damn words. I'm going to write them all here, so you don't have to. Ever.

Suppressors trap the expanding gas that leaves the barrel along with the bullet. The suppressor can't catch it all, but what does escape is negligible.

1911 with suppressor

What the suppressor can't silence is the action of the gun. You know, that dramatically loud sound the gun makes when the hero locks and loads before the final battle. That sound.

To be as silent as physically possible, your assassin or operator must do two things:

First, prevent the gun from cycling. This is more easily done in firearms without a hammer.

Simply have your character apply firm pressure to the back of the slide when they pull the trigger. This will give them one perfectly quiet shot.

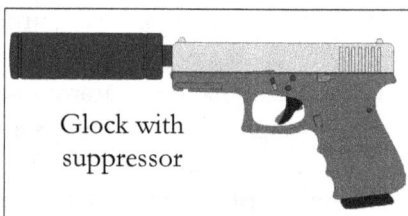

Glock with suppressor

This action has one main drawback. The gun must be manually cycled before it will be able to fire again. The secondary con is that, depending on how your character restricts the slide, it can sting a bit.

I did this once just to see how it worked. I fired a Glock with my firing thumb behind the slide. I thought I broke it.

On the other hand, your assassin could use a suppressed revolver to commit their crime. I know:

"Putting a suppressor on a revolver is stupid. The gas comes out between the cylinder and the barrel. Duh!"

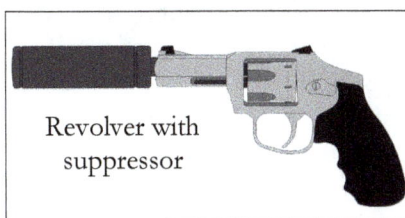

Revolver with suppressor

The fact is that many modern revolvers have tight tolerances between the cylinder and the barrel, so very little gas escapes from that gap. A suppressor is still beneficial because it would reduce a great deal of noise. However, the shot still won't be completely quiet. But, hey, some noise reduction is better than none.

You would probably catch a bunch of heat if you went with a silenced revolver unless you spend a lot of words explaining your character's rationale. So, best if you don't do this, but it could be done.

If your sniper is using a suppressed rifle, they probably don't have to worry about this. They are a good way off. What they do need to worry about is the crack of the round.

To meet this challenge, your assassin will use sub-sonic ammunition. The speed of sound is 1,125 feet per second. Any bullet with a velocity higher than that gives off a crack when it passes by.

Pistol Assassin: If the sound of the pistol cycling is removed, and the supersonic crack of the round is eliminated, you will have a true *Pfft* moment.

Rifle Assassin: The rifle assassin may not need to worry about sub-sonic ammunition. What they are mainly worried about is the report of the shot. That's what gives people the direction the shot was fired from.

If your shooter is 400 yards away and firing an unsuppressed rifle, witnesses will hear the crack of the round, followed by the report of the shot one second later. At less than 400 yards, multiple shots will sound like two or more weapons firing as the crack of the rounds overlaps with the reports of the rifle. Although it's nothing like experiencing this in reality, you can clearly hear the difference in videos from the internet.

If your shooter is firing a suppressed rifle, all witnesses will be able to attest to is the sound of the cracks – they won't be able to determine the direction the shots came from.

Video: https://www.youtube.com/watch?v=8HdgyoGdyV0

Your rifle assassin has a few things to do before they can just slap on the suppressor and go to work.

Suppressors use internal baffles to trap the expanding gas of the shot. These baffles create uneven gas pressures along the bullet's flight path inside the suppressor.

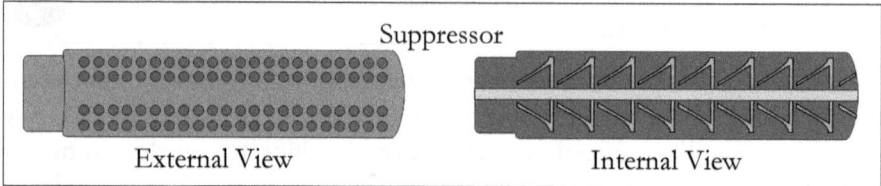

Suppressor

External View   Internal View

These uneven pressures alter the energy of the bullet when it leaves the barrel, which alters trajectory. Your pistol assassin may be close enough that this isn't an issue, but it's a big deal for the rifle assassin. They will need to zero the rifle with the suppressor attached before they can confidently perform their mission.

Suppressors are not durable items. Manufactured suppressors require maintenance to extend their lifespan.

Homemade/Improvised Suppressors
Your assassin may not have access to a machine shop to build a quality suppressor, but that doesn't mean they can't perform a little DIY. All you need is a tube on the end of the barrel filled with a baffling material.

I'll not provide any homemade suppressor plans here, sorry … lawyers.

Good news! The internet is full of ideas to solve your improvised suppressor needs. Have fun.

Note: If you're like me, your internet search history is an FBI nightmare. Not that they're monitoring or anything. Damnit, I just sparked my own natural paranoia.

Shotgun Ammunition

Shotgun cartridges are called shells. Externally, they're all similar. They consist of a plastic hull and a brass or steel base.

Shells with a longer base are called high-brass shells and generally contain more propellant, similar to magnum and +P ammunition. In fact, some shotgun ammo carries the magnum appellation. However, magnum shells may not be of high-brass construction.

Shotgun calibers are expressed in gauge. Why gauge? Because firearms manufacturers hate you and want to make your life difficult.

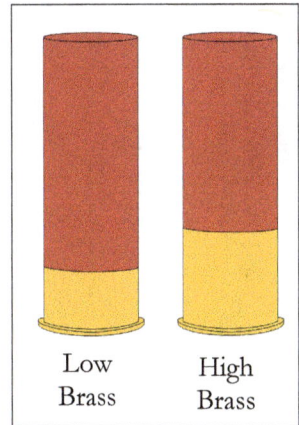

Low Brass   High Brass

Seriously though, this goes way back. In the early days of firearms people used to make their own projectiles – picture Mel Gibson melting down lead soldiers in *The Patriot*.

Back then, people could buy lead by the pound. If you bought one pound of lead and could make 12 equally sized balls to fit the barrel, you had a 12-gauge gun. The lower the number, the bigger the diameter. As you can imagine, shotgun gauge has nothing to do with how they measure pipe or wire.

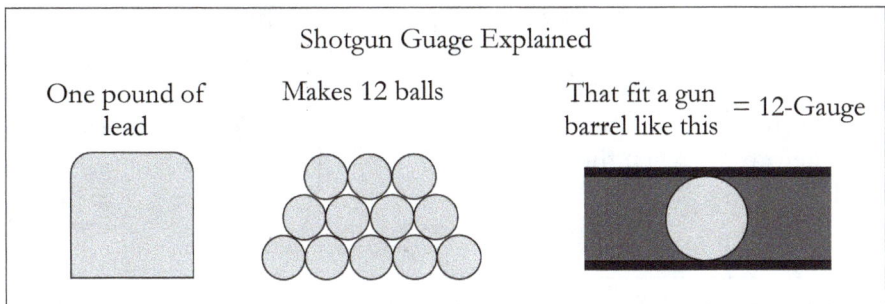

Shotgun Guage Explained

One pound of lead   Makes 12 balls   That fit a gun barrel like this = 12-Gauge

The measurement became the standard. It's stupid, and why they didn't change it when everyone was switching over to cartridges is beyond me. Okay, it's a standard we're stuck with, I get it ...

BUT NOOOOOO, they had to invent the .410-bore. It's the only shotgun size measured by caliber.

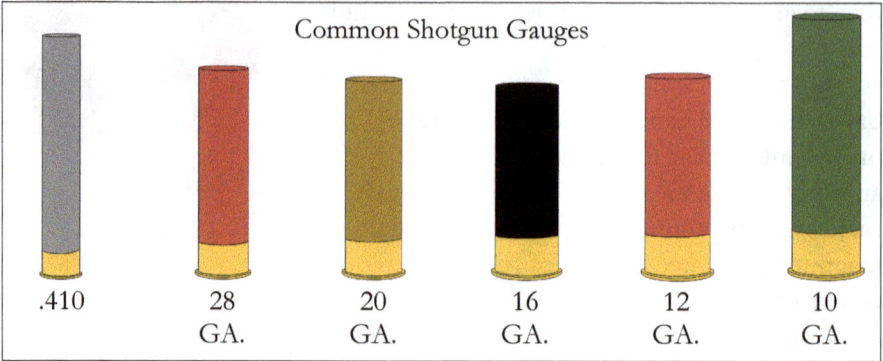

Common Shotgun Gauges

.410   28 GA.   20 GA.   16 GA.   12 GA.   10 GA.

(See page 254 for rules on writing shotgun ammunition.)

Shotshell Construction

The construction of the shotshell is different than that of rifle or handgun ammunition.

The base of a shotshell works just like any other cartridge. It contains a centerfire primer and houses most of the propellant.

The hull contains the wadding and shot.

Wadding is a plastic cup that contains the shot and protects it from deforming when the shell is fired.

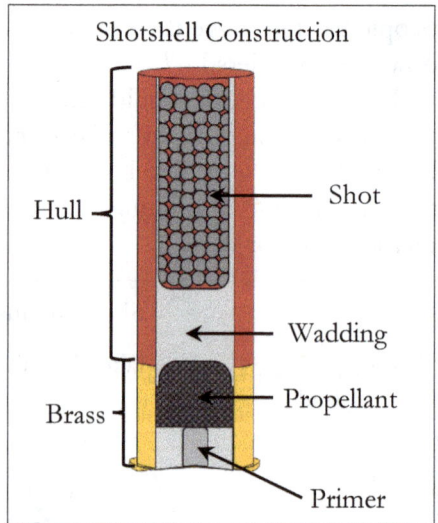

Shotshell Construction

Hull

Brass

Shot

Wadding

Propellant

Primer

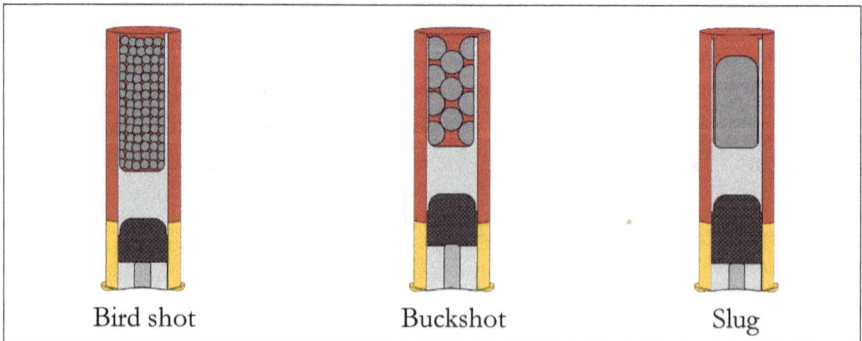

Bird shot   Buckshot   Slug

Shot Sizes

Shot takes the form of pellets or balls usually made of lead, steel, or some other material.

Because of the threat of contamination, the use of lead shot is often restricted for hunting to protect the environment.

Shot comes in a tremendous number of sizes and is generally classified as bird or buck. They increase in size as the number goes down.

Bird Shot									
Shot Size	9	8.5	8	7.5	6	5	4	2	1
Diameter in inches	.08	.085	.09	.095	.11	.12	.13	.15	.16

Bird Shot cont.				
Shot Size	BB	BBB	T	F
Dia in inches	.18	.19	.20	.22

Once the size of the shot reaches .24 inches, it's considered buckshot, and the numbering begins again.

Buck Shot								
Shot Size	4	3	2	1	0	00	000	0000
Dia in inches	.24	.25	.27	.30	.32	.33	.36	.39

Obviously, the smaller the shot, the more of it you can fit in the shell. Each shot size has its own applications and some overlap. For example, if you want to hunt small birds like pheasant or quail, you'd use smaller shot, so you don't destroy the meat. Smaller shot is also used for shooting sports like trap or clay.

Larger fowl require larger shot to kill the bird.

Buckshot and slugs are used to hunt larger game like deer, moose, and humans.

Thus, shotgun ammunition needs to convey its size and the load of the shell:

12-gauge #6 Bird // 12-gauge 00 Buck // 12-gauge Slug

Shotshell Appellations

So, if you compare the format for shotgun ammunition to rifle and pistol ammo, you can see the format is similar. Guage = caliber, and the appellation denotes shot size instead of chamber information.

.45-caliber ACP   //   12-gauge 00 Buck

However, in conversation, shotgun ammo is treated differently than rifle and pistol ammunition. With caliber, the appellation is often dropped: .45-caliber or *forty-five*.

Question: *What ya shootin?*

Answer: *Forty-five.*

With shotgun ammunition, once the gauge of the shotgun is established, it's the gauge that's often dropped instead of the appellation.

Question: *What ya shootin?*

Answer: *Double-ought buck.*

The information that is important to the conversation is either the size of the bullet (rifles/pistols) or the size of the shot (shotguns).

You may be wondering why hunters use buckshot and slugs when a rifle is so much more effective.

Some hunting areas restrict the type of ammunition used because of their proximity to population centers. Rifle rounds have tremendous range, whereas shotgun shells don't.

A .270-caliber rifle can effectively engage targets at 300-500 yards, while 12-gauge slugs are only effective to about 100 yards.

Patterning

The spread of shot is called pattern. The closer to the target, the tighter the shot is when it hits.

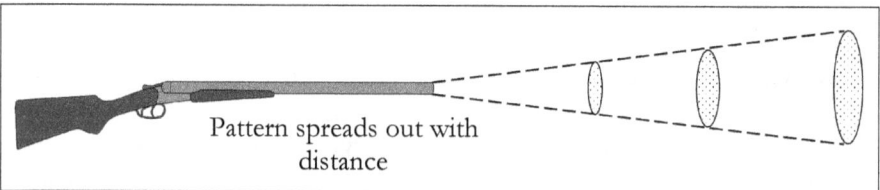

Pattern spreads out with distance

Patterning a gun is what hunters do to see where their shot goes at various distances. Shooters should use the same ammunition to hunt as they did when they patterned the gun.

There are devices called chokes that hunters and sportsmen mount in the barrel of shotguns to reduce the spread of the pattern. They do so by reducing the diameter of the barrel at the muzzle.

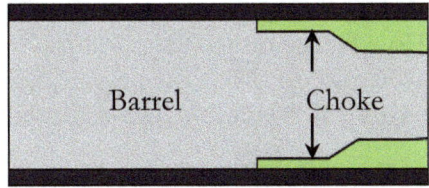

These chokes can also be offset to make the gun pattern left or right. This aids in sports like skeet shooting.

Rifled chokes are used to impart spin for shooting slugs. This stabilizes the projectile while in flight and makes it more accurate.

Other Shotgun Ammo

Because the shotgun is a smooth-bore firearm, you can fire almost anything through it. Ammunition types progress from less-than-lethal options such as rock salt and beanbag shells to pyrotechnics and mixed projectile options. There's even a Taser option.

Note that I said, "Less than lethal." This is an important distinction. Nothing fired from a gun is guaranteed to be non-lethal.

Range

How far the projectile(s) will travel is determined by several factors. However, for most the biggest factor is barrel length. The longer the barrel, the further the projectile will go.

Maximum Range is how far the projectile is able to travel when fired at the optimal angle, which is 45°. A 9mm bullet fired from a handgun, at the proper angle, can potentially travel three miles. Of course, it will have lost most of its energy and will do relatively little damage when it gets there. On top of that, what it hits is completely left to chance.

Maximum Effective Range is the maximum distance at which a firearm may be expected to be accurate and achieve the desired effect.

That's a vague definition, but good enough for us. What we're concerned with is what average engagement distances are for a given class of firearm. The table below lists average ranges.

Type	Range
Pistol	25 – 50 yards
Rifle	800 – 1000 yards
Shotgun (Bird shot)	50 yards
Shotgun (Buck shot)	50 yards
Shotgun (Slug)	50 – 100 yards

These are round numbers. Ammunition type, caliber, and barrel length will adjust the distances for each class. Obviously, a derringer with a two-and-a-half-inch barrel doesn't have nearly the range of a pistol, of the same caliber, with a four-and-a-half-inch barrel. The same applies to rifles.

Another factor to consider is the skill of the shooter. A rifle with the right ammo may have the *potential* to hit a target at a thousand yards, but is the character up to the task?

Thankfully, luck counts too.

Malfunctions

Scenario: The hero is in a shootout. He's armed with either a pistol or a rifle.

He had the villain in his sights. Time to end this, he thought, as he pulled the trigger. Nothing, the gun jammed.

All this work to pick the perfect gun, and then you go and jam it.

First off, there's nothing wrong with the passage above. My problem is with the word jam. I find it to be cliché. A deus ex machina used to take away a character's weapon at the worst possible moment.

I'm not saying don't use the word. It's common and people understand it to mean the firearm has a problem. I certainly don't expect you to stop in the middle of an action scene to describe the malfunction. That would be silly and goes against all my advice and good writing practices.

Merriam-Webster: jam definition b: "to become unworkable when a movable part becomes blocked or stuck." The example sentence is, no shit: *The gun jammed.* I LOLed.

As a reader, retired firearms instructor, and gun owner with thousands of hours of range time, my first question is always, "Why did it jam?"

In firearms lingo, we call them stoppages. Somebody somewhere classified them at some point, just like carry configurations, but I'm not going to explain stoppages in that format. I'm going to keep this as simple as possible. A stoppage is some interruption of the firing cycle:

Fire – Extract – Eject – Load – Fire

Stoppages are caused by one of the following reasons:

Maintenance: When poorly maintained, carbon buildup can foul the operation of the weapon. Additionally, lubricants can attract dust and dirt. So even if a weapon has little carbon buildup, it will still need occasional cleaning when exposed to the elements.

I once saw an AR that was seized up so badly, we needed to use a hammer and punch to get it open. The cause, carbon buildup. I've stress-tested guns before. I fired a bunch of rounds through them without cleaning, just to see how long it would take until they quit working properly.

When I finally cleaned those guns, they were a mess. However, they were clean compared to the one we had to beat on to disassemble.

Obstruction: Foreign matter like clothing, gear, or environmental detritus can be introduced to the weapon. This happens sometimes with pistols when the slide is prevented from cycling fully.

Ammunition: Occasionally, you get a bad round. Most often, this results in a simple misfire. Sometimes it can be worse, especially with reloaded casings.

My bad ammo story concerns a primer that popped out of the casing before the casing was ejected. That little bit of metal dropped down into the hammer and trigger mechanism of my M4. It took disassembly to figure out what was going on. I've only experienced this once.

Operator Error: Reassembly errors do happen. Put the wrong part in the wrong place, and the gun won't function properly. More often, this is an obstruction issue when cloth or cotton lodges unnoticed during cleaning.

We had an M4 on the range that simply wouldn't fire. Upon disassembly, we discovered part of a Q-tip was lodged in the bolt blocking the firing pin.

Magazine: If the magazine isn't functioning properly, the gun will experience feed-related stoppages. Magazines need to be cleaned too. If the follower becomes jammed, the next round won't be available to the bolt.

Additionally, if the follower is worn, the bolt won't catch on it and lock back when the last round is fired.

Mechanical Failure: A part of the gun breaks. This is the most uncommon malfunction. Yet it happened to me once.

To be fair, the gun was old. I mean really old. It was a refurbished Beretta 92 from the New Orleans Police Department. A friend of mine had bought it years before and fired a ton of rounds through it. Then I bought it from him and fired even more.

An internal part finally broke. There was no telling how many rounds that gun fired before the part failed.

We can see how rare mechanical failure is by looking at how firearms are tested before they're issued to service members or released for public sale.

- When the M1911 was field tested for the US Army, it was fired six thousand times nonstop. There was no rest for the gun and soldiers fired in shifts. When the gun became too hot, they dunked it in a bucket of water and resumed firing.

- H&K P30 was fired ninety thousand times during testing. The manufacturer performed this test with several sample guns and reported no major parts breakage in any of them.

To put those numbers in perspective, a typical range day for me equals two to four hundred rounds fired in the span of a couple hours.

Using the largest number, four hundred, I'd need to go to the range every day for four and a half months to equal the number of rounds H&K used to test its P30.

Interlude: Immediate Action Drills

Militaries, law enforcement agencies, and experienced shooters know that stoppages happen. Soldiers and police officers are given simple drills to deal with them. As a result, unless there is a catastrophic malfunction, most stoppages can be cleared in less than a minute.

As we used to say on the range, "You have the rest of your life to fix your problem." Black humor is the best humor, and shooters always got the message.

Automatic Rifle
S.P.O.R.T.S.

Anyone who's received professional or institutional training for the AR variant rifle is familiar with this acronym. This drill is the first step in reducing all stoppages. When the AR fails to fire:

S: Slap the magazine to ensure it's seated.

P: Pull back on the charging handle to eject the questionable round.

O: Observe that a round of ammunition ejects. This may or may not happen depending on the stoppage.

R: Release the charging handle, allowing the bolt to go forward and chamber a new round.

T: Tap the forward assist.

S: Squeeze the trigger. This step assumes you are actively engaging targets.

For automatic rifles without a forward assist, the drill is the same without the *tap the forward assist* part. In the coming pages, for rifles without a forward assist, if you see S.P.O.R.T.S., substitute this drill.

S: Slap the magazine to ensure it's seated.

P: Pull back on the charging handle to eject the questionable round.

O: Observe that a round of ammunition ejects. This may or may not happen depending on the stoppage.

R: Release the charging handle, allowing the bolt to go forward and chamber a new round.

S: Squeeze the trigger. This step assumes you are actively engaging targets.

The second drill is for a malfunction called a double feed. This is when two rounds are trying to chamber at the same time or brass fails to eject properly. This problem can't be cleared by S.P.O.R.T.S.-type immediate action. This next drill applies to all automatic rifles.

Rip – Roll – Rack – Reload

Rip out the magazine. This will strip one of the offending rounds from the magazine.

Roll the weapon so the ejection port is facing down.

Rack the charging handle twice. The first rack will allow loose rounds to fall free. The second rack will clear any round still in the chamber.

Reload the gun and get back to work.

Semi-Automatic Pistol

The procedures for the pistol are similar. However, the shooter is manipulating the slide, in most cases, rather than the bolt. The first drill is similar to S.P.O.R.T.S. and is the first action taken when facing a stoppage.

Tap – Rack – Bang

Tap up on the magazine to ensure it's seated.

Rack the slide, cycling the action.

Bang. Fire another round. Again, the terminology assumes you are in a firefight.

Semi-auto pistols can also encounter the double feed. The drill is virtually the same as the rifle.

Rip – Roll – Rack – Reload

Rip out the magazine. This will strip one of the offending rounds from the magazine.

Roll the weapon so the ejection port is facing down.

Rack the slide twice. The first rack will allow loose rounds to fall free. The second rack will clear any round still in the chamber.

Reload the gun and get back to work.

Addressing Malfunctions

So, we know what causes malfunctions and we know how to deal with them. In the following pages, I'll present stoppages from the most common to the least. I'll be doing page breaks to maintain continuity for each stoppage, and each will be addressed in the following format:

Title – Likely cause and explanation of the stoppage.

Symptoms – What the shooter experiences.

Remedy – Appropriate action by firearm type if the stoppage applies.

It should be noted that these malfunctions are typically experienced in the automatics, or detachable, magazine fed firearms.

It's rare for modern, manually operated firearms with internal magazines to experience most of these malfunctions. However, all firearms are subject to faulty ammunition.

Failure to Fire

This is one of the most common stoppages. It can be caused by bad ammunition or a light primer strike. Of course, it could be that your character failed to chamber a round. This is really embarrassing when it happens. There's nothing louder on the battlefield than: *CLICK.*

Symptoms: Click but no bang.

Remedy:

Manually Operated Rifle: Cycle the rifle's action.

Single-Action Revolver: Cock the hammer and attempt to fire the next chamber.

Double-Action Revolvers: Pull the trigger and attempt to fire the next chamber.

Pistol: Tap – Rack – Bang

Rifle: S.P.O.R.T.S

Failure to Feed

This stoppage presents as failure to fire. However, it's a completely different problem. It usually occurs because of faulty magazines.

In automatic pistols, this stoppage is sometimes caused by the operator not controlling the weapon properly. Known as limp-wristing, the shooter allows the gun to jump in their hand. This results in the gun not cycling fully, and the lead round isn't picked up by the slide.

Or it could be that the magazine is empty, which the operator should be able to determine during the execution of the drill.

Symptoms: Click but no bang.

Remedy:
    Pistol: Tap – Rack – Bang

    Rifle: S.P.O.R.T.S

Failure to Eject

Otherwise known as a stovepipe. Dirty guns or not enough lubrication are usually the culprits here. Also, limp-wristing in the automatic handguns.

Symptoms: The operator has a mushy trigger, and there's no click. This is cause for immediate inspection of the gun. The character will notice that there's brass sticking out of the ejection port.

Remedy:
    Pistol:
- Place the non-firing hand on top of the slide forward of the brass.
- Sweep the hand backward. This will pop the brass out.

    Rifle:
- Place a hand on the slide of the receiver forward of the brass.
- Sweep the hand backward. This will pop the brass out.

Double Feed

This is when two rounds are trying to chamber at the same time or brass fails to eject properly.

Note: I've not experienced this before in a manually operated firearm. However, it is possible.

Symptoms: The operator experiences a mushy trigger and no click. Again, this is cause for immediate inspection of the gun. When there's no brass sticking out of the ejection port, a closer look will reveal the double feed.

- Experienced operators will proceed directly to the correct drill.
- Novices will attempt the primary drills Tap – Rack – Bang or S.P.O.R.T.S. Neither of these will work.

Remedy:

Semi-Autos: Remedy: Rip – Roll – Rack – Reload

Manually Operated Rifles:

- Open the action fully.
- Tilt the firearm and shake out the offending cartridges.
- Close the action then re-rack.

Brass Over Bolt

This is specific to AR-type rifles. An expended casing or live round becomes lodged over the bolt. A well-trained operator can clear this stoppage in less than a minute.

This is a catastrophic stoppage for the untrained.

Symptoms: The operator will experience a mushy trigger, no click, and the charging handle will be stuck. Visual inspection will reveal the stoppage.

Remedy:
- Remove the magazine.
- Grasp the charging handle and, while applying backward pressure on it, bang the butt of the weapon on the ground. This will retract the bolt far enough for the brass to fall free in the next step.
- Reaching up through the magazine well, use a thumb or finger to apply pressure to the bolt face. Then, using the other hand, tap the charging handle forward. The offending brass will fall free.
- At this point, regrasp the charging handle and retract the bolt. Don't try to pull your finger out without performing that last step. The bolt will cut your finger. Full disclosure: I learned that lesson the hard way.

Failure to Extract

This can be caused by a bad ammunition casing that tears, causing the brass to become lodged in the firearm, or a bad extractor that is not grabbing the rim of the casing. This is a good malfunction to use to kill your character's gun.

Symptoms: The operator feels a mushy or soft trigger. No click. This stoppage presents as a double feed.

Note: I've never personally experienced this malfunction.

Remedy:
Semi-Autos:
- Rip – Roll – Rack – Reload. Only one round will fall free.
- The operator will induce another double feed when they try to reload.
- Upon further inspection, the brass stuck in the chamber will reveal itself.
- With the action open, the operator slides a rod down the barrel and punches out the offending casing.
- If it's too badly stuck, an armorer may need to be consulted.

Manually Operated Rifles:
- Open the action fully.
- Tilt the firearm and shake out the offending cartridges. Only one round will fall free.
- Close the action, then re-rack.
- The operator will induce another double feed when they try to reload.
- Upon further inspection, the brass stuck in the chamber will reveal itself.
- With the action open, the operator slides a rod down the barrel and punches out the offending casing.
- If it's too badly stuck, an armorer may need to be consulted.

Mechanical Failure

Cause unknown.

Symptoms: Either click with no bang, or mushy trigger and no click.

Remedy: The operator will attempt immediate action drills escalating from Tap – Rack – Bang/S.P.O.R.T.S (as per weapon type) to Rip – Rack – Roll – Reload in an effort to fix the problem. Final resolution will be determined by disassembly and inspection.

Taking out the Gun

Guns are machines, and machines break, always at the most inopportune time … like my snow blower this year.

The probability that your character experiences a catastrophic gun malfunction is fleetingly rare. Most stoppages are resolved quickly enough to not be an issue. That is, unless proximity comes into play. If your villain is within a reasonable distance, they can close that distance and attack the hero before they can fix the malfunction.

He had the villain in his sights. Time to end this, he thought as he pulled the trigger. Nothing happened. He franticly worked to correct the **malfunction/stoppage/jam**, *but the villain was upon him. As he was thrown to the ground, the gun slid away.*

Monologuing can happen after the initial disarming move, or you can proceed directly into the awesome fist/knife fight.

Note: All three terms (malfunction/stoppage/jam) are acceptable. Write what you're comfortable writing.

Let's refresh the scenario and add more options.

Scenario: The hero is in a shootout. His primary weapon is a rifle, and his secondary is a pistol.

He had the villain in his sights. Time to end this, he thought as he pulled the trigger. CLICK. The gun didn't fire.

With the addition of a secondary weapon, the professional, depending on proximity to the target, has two options:

Option One: The villain is far away, and the hero has time or cover. The hero would attempt to reduce the malfunction and get his primary weapon back in operation.

Option Two: The villain is close, and the hero has no time to fix the gun. The hero would immediately transition to his secondary weapon, the pistol. (Page 115)

A trained, experienced operator – military or law enforcement – will never discard a weapon by choice, especially an issued one. They will perform immediate action to the point that they realize the weapon is useless and retain it as long as possible.
- If the character has a secondary weapon, the rifle will be slung.
- A pistol will be holstered.
- If the hero doesn't have a secondary, the rifle can be used as a club.

My point is this: Modern firearms are designed not to malfunction. Ammunition is designed to fire. Malfunctions do happen but not as often as we are led to believe by fiction and media.

A well-timed jam is useful, but it's also an overused trope.

I recently went to the range with my cousin. I fired 400 rounds in about an hour. I wasn't even pushing the gun that hard. In that time, and all those rounds, I didn't have a single stoppage. If I did:

- Failure to fire = Tap – Rack – Bang
- Double feed = Rip – Roll – Rack – Reload

On top of that, in over thirty years of working with firearms, I've personally experienced three incidents where the weapon had to be disassembled to be fixed.

If firearms failed in real life at the same rate they do in fiction, armies would use swords. As writers, we can find more imaginative ways to take away the gun other than, "The gun jammed."

So, given the fact that your well-armed, well-trained character has options when their firearms fail, *you* have four options to take them away.

Option One: Run your character out of ammo.

Running your character dry can be difficult. Weapon choice can help you. If you choose a firearm with high-capacity magazines, you may be screwed here. However, the nature of the fight may come into play and help. After all, it's difficult to hit a moving target. Especially if you're moving and dodging too.

Option Two 2: Damage the gun.

People get shot; guns get shot. It's an old trope to shoot the gun out of someone's hand, but it does happen. Police and military are taught to shoot at the center of mass of the target offered. Where is the gun held? Pretty close to center mass.

When I trained with Simunitions, a training round that shoots a plastic projectile loaded with paint, I got hit in the hands and the gun all the time. I also used to shoot my opponent's guns as a teaching point. If the gun gets shot, it's probably not going to work. But let's look at a real-life example.

Miami Shootout, 1986

During the firefight:
- Special Agent Jerry Dove's S&W Model 459 9mm pistol was hit by a bullet and rendered inoperative.
- Special Agent Gordon McNeill was struck in the hand. This wound, and blood in the chambers of his revolver, made reloading impossible.

Option 3: Use calamity.

I mentioned getting shot in the hands. Also during the Miami Shootout, Platt – one of the bad guys – was shot through the forearm. This broke his radius bone and caused him to drop his pistol.

Option 4: Time and distance.

Your character could have all the spare magazines in the world – it still takes time to reload. They could be proficient in clearing a stoppage or transitioning to another firearm, but those things take time too. Place your combatants in close proximity and don't give them the gift of time.

In the chaotic moments of a gunfight, equipment gets dropped or left behind all the time. The best way to take the gun away from a character is to fully choreograph your fight. (Page 271)

Combatants, Choose your Weapons

I'm often asked, "Hey Chris, what's your favorite handgun?"

My reply, "The one in my hand when I need it."

It sounds like a smart-ass answer, but it's true. As it happens, I almost experienced a home invasion. I ended up holding the guy at gunpoint with my backup pistol, a Kahr K9. All my other guns were in my car in the garage in preparation for the next day's training on the gun range. My favorite gun that night was the Kahr.

Fortunately, or unfortunately for your characters, you get to choose. So how do you pick what's right? Before you decide, consider these factors:

Purpose

What is your character going to do with the firearm? Firearms are designed to perform a certain role. You can hunt deer with an AK47, but it's not the best rifle for the task. Conversely, there are handguns that are specifically designed for hunting that you can't just tuck in your belt.

Pfeifer Zeliska .600 Nitro Express Revolver

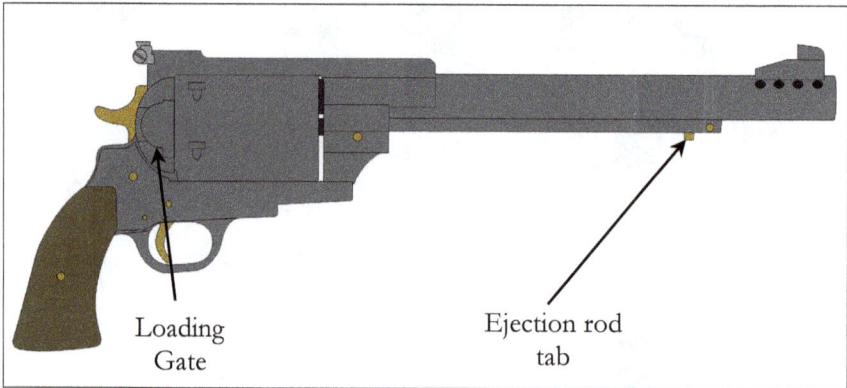

Loading Gate

Ejection rod tab

I couldn't resist this oddball. The Pfeifer Zeliska .600 Nitro Express was designed to hunt elephants and other dangerous game. It's a five-shot single-action revolver that operates in the same manner as the Colt Peacemaker.

This ridiculous revolver is twenty-two inches long, almost as long as the M4 carbine. Not your everyday carry revolver.

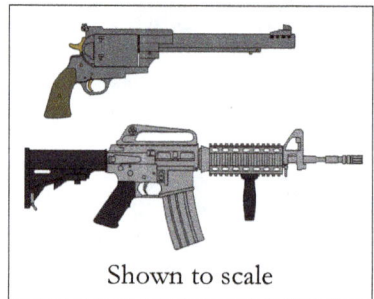

Shown to scale

You want the right firearm for the right job.

Longarms

Do you want a rifle or a shotgun? Understand that each has its own applications and effects. For long-range work, a rifle would be best. At close range, things become muddier.

Shotguns work well inside fifty yards/meters. If you go past that, accuracy becomes unreliable.

Rifles work well at all ranges and generally have a higher magazine capacity.

Handguns
Coolness Factor

Let's say you're not restricted by agency or time period. Now, you want something that will really embody who your character is. Harry Callahan had his .44 Magnum. Bruce Willis used twin .45s in *Last Man Standing*. The gun is an accessory that tells us who your character is.

If the name of the gun is cool, so is the character. There are plenty of guns out there with super cool names.

- Kimber R7 MAKO
- Desert Eagle .50
- Springfield Hellcat

Size and Concealability

I talked a little about printing in the holsters section. Obviously, the smaller the gun, the easier it is to conceal. The early James Bond carried a Walther PPK, which is considered a sub-compact pistol. This is not a random designation. Handguns are also classified by size:

- Full size = > 4.5" barrel
- Compact = 3.5 to 4.5" barrel
- Sub-compact = 3 to 3.5" barrel
- Pocket = < 3" barrel

Grip size comes into play as well. Full-size and compact pistols fill the hand, while sub-compacts and pocket-sized don't.

Other Factors

Function

I can't emphasize this enough. If you're going write about how your character manipulates the gun, you must choose one that has that functional part. You can't cock the hammer on a DAO revolver, and you can't engage the safety on a Glock.

Pick a gun with the features that suit your scene.

Organizational Issue

If your character is part of an agency or military, you are restricted to the firearms that organization uses. Obviously, a US soldier is not going to be issued an AK 47. They'll be equipped with what the military gives them. This brings us to ...

Time Period

Sticking with the military theme, A soldier prior to 1969 was issued the M14 rifle. It was replaced by the M16 variants in 1964, which, in turn, were replaced by the M4. Full conversion to the M4 occurred around 2005. Currently, the US military is working to convert to the new rifle, the XM5. Time period matters.

Police and federal agencies also periodically adopt new weapons. These adaptations often occur because of major shooting incidents. The FBI and several other law enforcement agencies traded in their old service revolvers for semi-autos after the Miami Shootout in 1986.

Similarly, the North Hollywood Shootout, 1997, convinced police agencies across the country to rethink their cruiser longarm. Thus, many officers now have an M4 or other rifle in the trunk rather than a shotgun.

Finally, you have to consider whether the gun was even in existence for the time period. Unless your western includes time travel, your cowboy won't be armed with an Uzi. Okay, extreme example. Your WW1 German soldier won't be armed with a Walther PP.

POV

I was recently asked what type of gun should be used for a scene in a story that was written from a mugging victim's POV, and the gun never makes another appearance in the story.

There are times when naming the gun doesn't matter. If it never makes another appearance or doesn't impact the plot or action of the scene, just call it a pistol or revolver, and move on.

Here's another one I get sometimes. "How would an experienced operator know what kind of gun was pointed at them?"

An operator wouldn't care. The only thing they would be concerned with is how to get the gun away from the bad guy, how to draw their own gun and shoot the bad guy, or whether they can run away without getting shot.

But let's step back for a minute and examine the problem. All the character sees is the muzzle and the outline of the slide or frame. That's precious little to make an identification from. True, there are some distinctive characteristics to certain guns. However, with the proliferation of different makes and models, too many are too similar for a definitive ID.

The best your character can do is identify whether the handgun is an automatic or a revolver.

The type of gun may be important to the story, but weigh reader knowledge against character knowledge, or the ability of the character to observe. This will inform you on whether you should identify the gun in that passage.

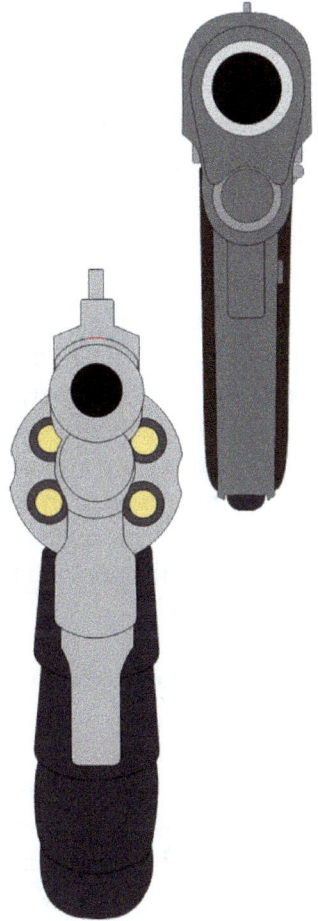

Choosing a Firearm

Recently, my wife wanted to purchase a pistol for security when I travel. Now, I must say, I have a few guns. As a firearms instructor, I need to have a proper selection to give proper instruction. You can't show up to training and say, "Hey dude, can I borrow that for a minute?" It would ruin my credibility.

But none of my guns were the right fit for my wife. She's a tiny lady, so the shotgun and the 1911 were off the table. She doesn't like semi-autos. She wanted a revolver. So, when we went to the gun store, she used many of the criteria listed below to make her selection.

If your character's choice in firearm is not restricted by an agency, they should use the following criteria when choosing a gun:

Comfort

Comfort is one of those ineffable qualities specific to the character. I, personally, don't find the 1911 to be a comfortable gun to shoot, not because of the caliber, but because of the grip safety. Other people love the way it feels and shoots.

Comfort is composed of a multitude of factors. One of the biggest is grip angle in relation to the barrel. There's a lot going on here, but basically, if the angle is complimentary to the shooter's hand and arm structure, the gun will feel more natural to them. It will also be easier to aim because they won't need to cant their wrist to line up the sights on a level target.

The character should ask themself if the gun feels good in their hand. If the gun is uncomfortable to hold, it'll be uncomfortable to shoot. If it's uncomfortable to shoot, the character won't train with it. If they don't train with it, they won't be proficient with it, and then the gun becomes a liability instead of an asset. This applies to all firearms, both shoulder fired and handgun.

Ease of Use

Can the character reach all the parts to make the gun work? I have tiny hands. With some pistols my thumb can't reach the slide release. This will rule them out for me.

Left-handed characters will need to look for guns that are either ambidextrous or can be converted for left-handed use. Many firearms are manufactured so magazine releases can be switched to the other side.

If the firearm isn't set up for this type of conversion, aftermarket parts are available to convert almost every function for any gun. However, availability may be restricted to the most common firearms.

In rifles and shotguns, can the character reach the trigger when the gun is braced against the shoulder?

Safety Features

Does the character want a pistol with a safety or without? The more stuff that's on a gun, the more the character must train with it to feel confident in its use.

Caliber

Is the firearm going to overpower the character? Will they be able to control the gun when they fire it? The gun may feel good in the hand, and the character may be able to manipulate it well, but if they can't control it when they shoot it, the gun is a liability.

Ego

Is the character selecting this gun because it's iconic? Does image play a factor? If the gun is a vanity purchase, any or all the above criteria could be used as a liability against said character.

Gun Choice Traps

Some guns are just not good for fiction. They conceal inherent traps for the writer. Let's look at our first offender, Glock.

Model versus Caliber

Glock names their gun models after the patent number the gun is manufactured under. This is a terrible idea, and it proves that Glock hates writers. For example:

- The Glock G22 is chambered in .40-caliber S&W.
- The Glock G40 is chambered in 10mm.
- The Glock G44 is chambered in .22LR

All of those model numbers are calibers, but not the caliber of the model. So, if you write:

He pulled his Glock 40.

Do you mean the G40 chambered in 10mm, or did you want the G22 chambered in .40-caliber S&W? When writing Glocks, be very careful when referring to model number and caliber. Especially if you're going to use a weapon tag or epithet. (Page 241)

Model versus Features

Another offender lurking in the page of this very book is Heckler and Koch. The P30 has two variants available on the market.

The P30 Variant 1 is a striker-fired pistol similar to a Glock. It has no hammer and no safety. Well, it does have the trigger safety, but you know that doesn't count. (Page 167)

The P30 Variant 3 is a DA/SA pistol with a de-cocking lever and a separate safety. But wait, there's more. The Variant 3 is also available without a safety.

So, you could be completely correct in referring to the safety on a H&K P30, but unless you've included the variant number, you've left yourself vulnerable to readers who are only aware of Variant 1.

H&K was at least kind enough to name the variants. Some guns use the same model number for their variants even though they have different features. The FN Model 1910 was available in both .380-caliber ACP and .32-caliber ACP, yet FN didn't feel the need to distinguish between the two variants.

Not all firearm manufacturers are as inconsiderate to writers as H&K. Most will denote their variants by adding numbers or letters after the base model. The Beretta 92 has seen a number of variants, and they are denoted by letters behind the model number. The original 92 didn't have the de-cocking safety. That modification was installed on the 92S.

The M16 also went through its evolution, culminating in the M16A3. Be careful with the M16. The A1 could select semi- or full-auto. The A2 model fired a three-round burst rather than full-auto. This was an attempt at regulating ammo consumption. The full-auto feature was re-instituted in the A3.

Nomenclature

The last big trap is nomenclature. As noted in the introduction, different manufacturers will sometimes use different terminology for a part that performs the same task: Glock – slide stop/Beretta – slide release.

This may not be a big deal, depending on how the sentence is constructed. Using a Glock:

He thumbed the slide release, chambering a round. INCORRECT

He released the slide, chambering a round. CORRECT

The first example is incorrect because of nomenclature. Will a Glock fan call me out? Hard to say. If the action and story are tight, I might get a pass. On the other hand, Glock aficionados prefer the slingshot method over the slide release method when performing emergency reloads. So, they might give me grief about technique instead. (Page 181)

The second sentence works with all nomenclatures and slide release techniques. I don't risk using the wrong term, and the action is vague enough that the reader can fill in the blanks with their own experience and preferred technique.

Best Advice:
- Avoid, if possible, guns with model numbers that are similar to a caliber, unless the caliber and the model number are the same.
- Avoid, if possible, guns with multiple configurations under the same model number.
- Perform a full diagnosis of the firearm and note discrepancies in nomenclature.

Cheat Sheets

If you're new to firearms, or have many to keep track of, I suggest creating a cheat sheet for each gun in your story. It should contain all the information you need for quick reference.

These are the minimum characteristics you need to identify:

Handguns
- Name: Manufacturer and model
- Type: Revolver/Semi-auto
- Capacity: In revolvers, it's the number of chambers in the cylinder. In semi-autos, you may see a number followed by +1. This means the magazine can be ejected and plussed back up to full capacity. Guns like the Grendel P10 and the Mauser C96 don't have a +1 option.
- Trigger Action: SA, DA/SA, DAO, Striker
- Safety: Yes/No. Note: Trigger safety doesn't count. (Page 167)
- Magazine/Cylinder Release: Note the location. Most mag releases are in the form of a button. Some are levers on the trigger guard, and still others are a latch at the bottom of the gun. Cylinder releases can be pushed forward, pulled back, or a button.
- De-Cocker: In DA/SA guns this may be the safety or a separate lever.
- Slide Release: Note nomenclature differences if possible. Not all pistols are equipped with this feature.

Long Guns
- Name: Manufacturer and model
- Action: Automatic/Bolt/Lever/Pump
- Magazine: Ejecting/Internal Box/Internal Tube
- Capacity: Specific to internal magazines
- Safety: Location. Type: Slide/Button/Selector
- Magazine Release: Note the location if applicable.
- Bolt Manipulation: Automatics. Note nomenclature.

The following examples are just a few ways to organize important information. I'll include blank versions in an annex. Our first example is what I call the baseball card, because it sends me back to the days of my youth. I'll use a Kimber .357.

Manufacturer:	Kimber	
Type and Action:	DAO Revolver	
Model:	K6s	
Caliber:	.357 MAG / .38 SPC	
Capacity:	6	
Size:	Full size = > 4.5" barrel	
	Compact = 3.5 to 4.5" barrel	
	Sub-compact = 3 to 3.5" barrel	
	Pocket = Less than 3"	
Safety:	NA	
Suggested Carry Condition	All chambers loaded	
Condition when empty	NA	
Clicks when empty*	Yes	
Gun Tags (See Page 241)	Kimber / Magnum / .357	

But maybe you want to track more information about the firearm. For this fuller version, I'll use my retirement gift, the SIG Sauer 1911.

Character:	Chris
Manufacturer:	Sig Sauer
Type and Action:	Single Action Pistol
Model:	1911
Caliber:	.45 ACP
Capacity:	8, +1

Size:	**<u>Full size = > 4.5" barrel</u>**
	Compact = 3.5 to 4.5" barrel
	Sub-compact = 3 to 3.5" barrel
	Pocket = Less than 3"
Safety:	Yes, lever on back of slide.
Suggested Carry Condition	Loaded, hammer back, on safe
Condition when empty	Slide locks to the rear
Clicks when empty*	No

Historical Data	
Dates of production	2004-Present
Military Use	1911 – Yes: Not this Manufacturer

Character data	
Character Carry Condition	Israeli style: Mag in, chamber empty, safety off
Character Carry Location	Shoulder Holster
Staging Location (Home)	Nightstand
Staging location (Car)	Glovebox
Staging Location (Office)	Upper right desk drawer
Gun Tags (See Page 241)	Sig / 1911 / .45
Notes and Identifying Features:	

Chris's nickname for the gun is Big Bertha. He received the gun as a retirement gift from his unit when he left the Army.

You may have noticed the asterisked entries. These are categories that I feel must be included in your cheat sheet.

The first, *Clicks when empty*, will help you avoid one of the most common errors in fiction. Not all guns click when empty. Especially the automatics. See page 86 (automatic rifles) and page 159 (semi-auto pistols) for more information on this error.

The second note is for gun tags. These are acceptable terms you can use in place of common nouns, like pistol and gun, without needing to say the firearm's full name. See page 241 for more advice on gun tags.

The cheat sheets are nice, but once you've become familiar with your stable of guns, you may only need a quick reference table to check a detail.

This next suggestion cuts things down to the barest of facts you'll need to keep from confusing your firearms.

Name / Features	Beretta 92	Grendel P 10	1911	Kimber K6s
Manufacturer	Beretta	Grendel	Sig Sauer	Kimber
Model	93fs	P 10	1911	K6s
Type	Semi-auto Pistol	Semi-auto Pistol	Semi-auto Pistol	Revolver
Caliber	9mm	.380 ACP	.45 ACP	.357 Mag / .38 SPC
Capacity	15 +1	10	7 +1	6
Trigger Action	DA/SA	DAO	SA	DAO
Safety	Yes	No	Yes	NA
Magazine Release	Yes	No	Yes	NA
De-cock	Y safety	No	No	NA
Slide release	Yes	Yes	Yes	NA
Clicks on empty*	No	No	No	Yes

Whether you use a full sheet for the gun each character uses, a baseball card, or a simple table, you should fill it out as you research the firearms that will go into your story.

Researching the Firearm

This may seem remedial. If you're reading this, you already know how to research the details you put in your stories. However, it can be surprisingly tricky to nail down the features of a firearm.

I start with Wikipedia for the basics. Let's use *Shooting Illustrated's* Handgun of the Year for 2005, The Rohrbaugh R9.

Wiki will tell you:

- Name: Rohrbaugh R9
- Type: Semi-automatic pistol
- Capacity: 6 +1
- Trigger Action: Not mentioned
- Safety: None
- Magazine Release: European style at the base of the grip
- De-Cocker: Not mentioned
- Slide Release: None

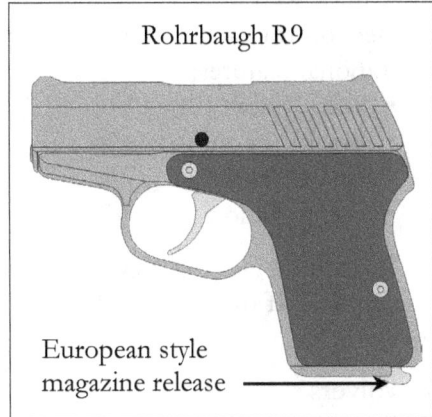

Rohrbaugh R9

European style magazine release ———▶

There's no mention of how the trigger operates. However, from the picture, we can guess that it's either striker-fired or double-action-only (DAO) because there's no visible hammer. Further reading confirms my guess. This gun is DAO.

The thing that gives me pause is the lack of a slide release. I couldn't find anything in the literature that explains why it's absent and what effect that has on someone firing the gun.

Eventually, I found, in a video, that the R9's slide doesn't lock back on an empty magazine. Your character will need to work the slide after reloading. That's the trap of this gun. It doesn't function in the typical way.

If you find the gun doesn't have expected features in the expected locations, research further. Don't assume anything.

I prefer to do image searches and perform my own visual inspection of the firearm to discern how it functions and what components a character can interact with. For me, this technique works about 99% of the time. However, oddballs like the Rohrbaugh sometimes trip me up.

Firearms of a certain type will have common features. If they're present, they'll be in expected locations. If the feature is not visible on either side of the gun, then the gun operates in an unexpected manner and could represent a trap for the writer, as shown with the R9 on the previous page.

The following pages list the expected locations of a firearm's operational features. When using this technique, be sure to check both sides of the gun. The part you're looking for could be hiding on the other side.

Long Guns

Manually operated long guns are much easier to diagnose because there have been so few design changes to them in the last seventy years. See the section for each type for the expected locations of the gun's operational features:

- Bolt Action: Page 80
- Break Action: Page 69
- Lever Action: Page 78
- Pump Action: Page 75

If the long gun has an ejecting magazine, it will be in the firearm's description. If the gun has this feature, a picture should be a dead giveaway.

Revolvers

Revolvers are also easy to diagnose from a picture. See the section for each type for the expected locations of the revolver's operational features.

- Single-Action Revolvers: Beginning on page 145
- Modern Revolver: Beginning on page 148

Machine Guns

Machine guns have fairly common anatomy. See the diagram of the M60 for the expected locations of the machine gun's operational features. (Page 100)

Derringers, semi-auto pistols, machine pistols, submachine guns, and select fire rifles are a little more difficult to diagnose from a picture because there are so many variations in their designs. Thus, I've developed a graphic for each to show the probable locations of their operational features. Submachine guns and select fire rifles are lumped together because they're similar in design.

Derringers

Expected locations of operational features.

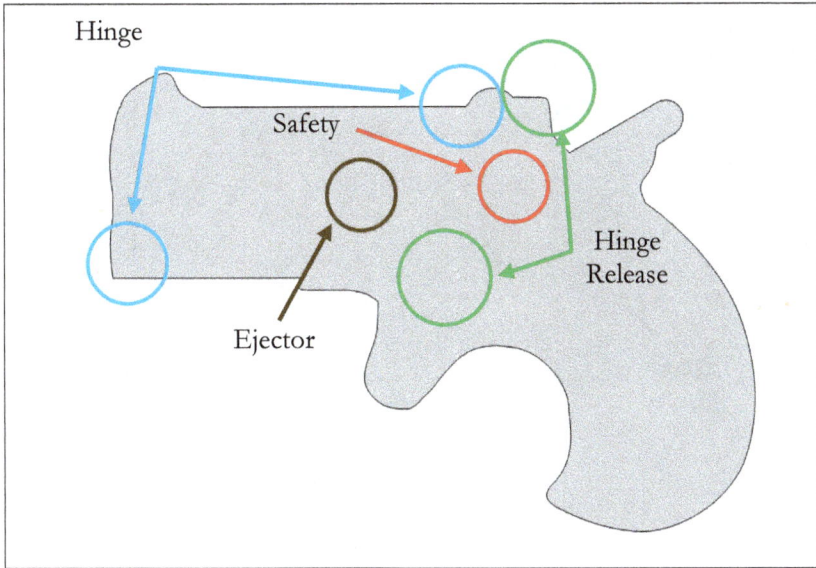

What to look for:

- **Hinges:** Hinge points may be obvious. However, they sometimes look like screws or pins. Hinges for two-barreled derringers are usually on top, while hinges on four-barreled models are usually on the bottom. Be careful of designs like the Sharps derringer. (Page 138)
- **Hinge Release:** The hinge or action release is usually found on the frame near the trigger and looks like a lever. In the COP .357, the release is combined into the rear sight and is located on top of the gun.
- **Ejector:** This is common in newer derringers and, if present, will be located on the side of the barrel assembly in the form of a tab.
- **Safety:** If present, it usually takes the form of a button.
- If a hammer is present, it suggests the gun is single action – most are. If there is no hammer, the firing pin rotates when the hammer is cocked. Beware! There are derringers with a double-action trigger.

Some derringers are equipped with a trigger guard.

Semi-Auto Pistols

Expected locations of operational features.

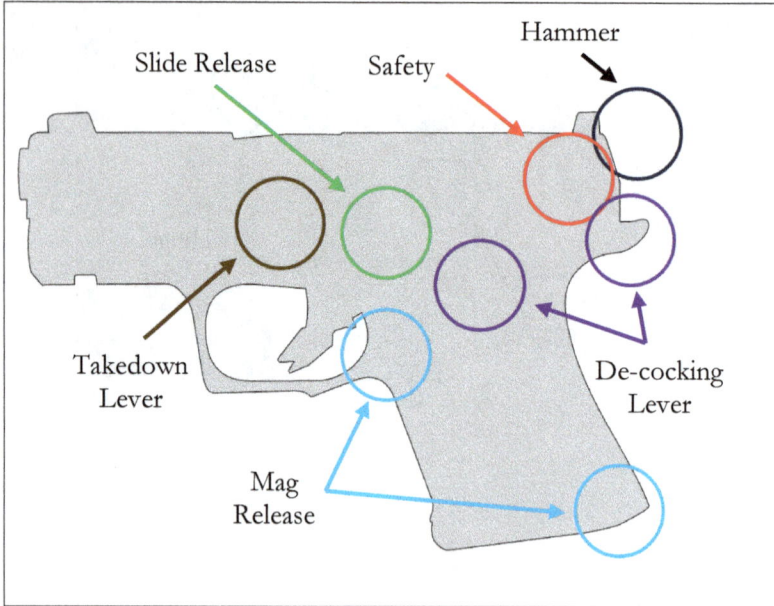

What to look for:

- **Hammer:** If the pistol has no visible hammer, you can safely assume it's striker-fired. Pistols with a hammer can be either single action or DA/SA.
- **Safety:** The safety will almost always take the form of a lever. Safe position is usually up. Fire may be denoted by a red dot or lettering.
- **Slide Release:** Almost always at the midpoint of the slide. Caution! Takedown levers, if present, are usually close to this location.
- **Takedown Lever:** If present, a takedown lever is used to disassemble the gun. They are usually located slightly forward of the gun's midpoint. Don't confuse this lever with the slide release
- **Mag Release:** Usually, a button or lever near the trigger. Can be a catch at the butt of the grip. This is known as European style.
- **De-Cocking Lever:** The de-cocking lever is usually part of the safety, but on guns like the Sig P226 and the H&K P30 Variant 3, this a separate component.

Machine Pistols

Expected locations of operational features.

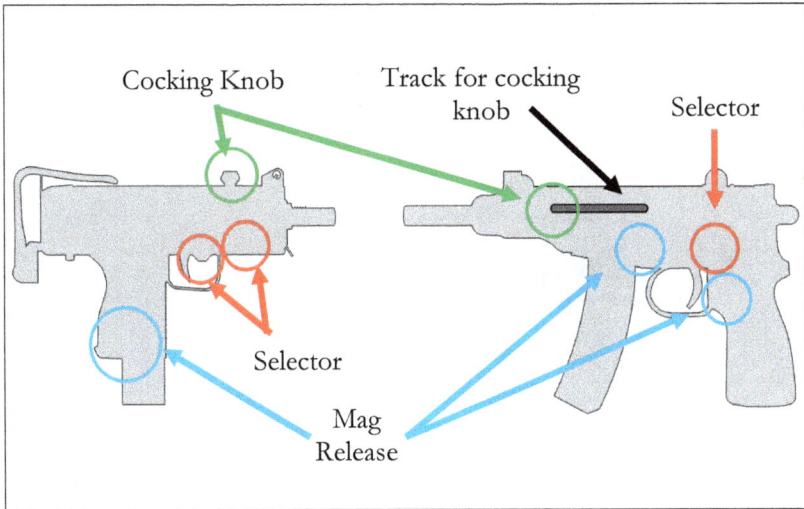

Machine pistols come in two designs. The first is like the Uzi, or Ingram MAC-10, where the magazine is located in the grip. The second is like the MP5 or Škorpion VZ61, where the firearm has the magazine located forward of the trigger.

What to look for:

- **Receiver Size:** Machine pistols are larger than most pistols. They need the extra room for sturdier bolt and spring assemblies.
- **Cocking Knob:** May be located on the top or side of the firearm. If it's on the side, there will be an obvious track in the receiver. This track indicates that the part at the forward end performs the cocking function.
- **Selector:** Because these firearms are designed to fire full auto, there will always be a selector. It will be located on the receiver, and it could take the form of a switch, lever, or slider.
- **Magazine Release:** The magazine release could be a button, lever, or catch.

Firearms of this type fire from either an open or closed bolt.

Submachine Guns and Select Fire Rifles
Expected locations of operational features.
(Barrel and stock removed for clarity)

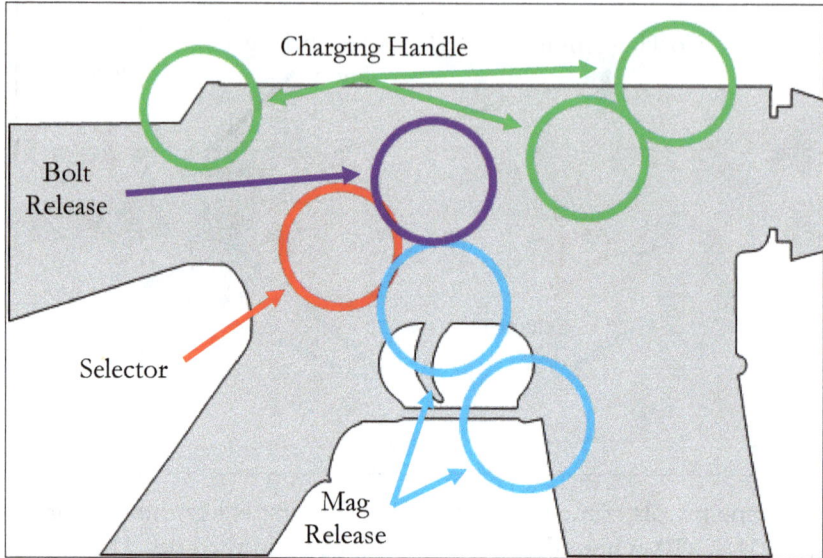

Charging Handle

Bolt
Release

Selector

Mag
Release

What to look for:
- **Charging Handle:** There will always be a way to manipulate the bolt. The charging handle on the AR series rifles is an anomaly. Most charging handles or cocking knobs are located toward the front of the receiver. It may be on top, on the same side as the ejection port, or on the opposite side of the gun.
- If the charging handle is on the side, there will be an obvious track for it to travel in.
- **Bolt Release:** The bolt usually locks to the rear on the final shot. After the fresh magazine is inserted, the shooter can depress the bolt release and chamber a round. If there is no obvious bolt release lever or button. The bolt is released by pulling back and releasing the charging handle.
- **Selector:** The selector is usually located near the trigger for easy manipulation. AK-style firearms will have an ungainly lever on the right side.
- **Mag Release**: Usually, a button or lever near the trigger. Can be a lever behind the magazine.

Wading through videos can be tedious and time-consuming, so I like to use pictures to identify a firearm's operational features. This technique works for me because I know guns and what to look for.

If your visual examination of images fails to reveal all operational components, search for a video review of the gun. But beware, some of these guys have no idea what they're talking about. I confirm everything with two or more sources.

If you find a gun you just have to use, but you can't find all the particulars on it, keep character interaction simple.

The character (loaded/ aimed/ fired/ cleared) the gun.

Gun Tags and Epithets

Congratulations! You've picked the perfect gun for your character. You know its features and how it functions. Now you can do things that writers who only use generic guns can't. You can use epithets and what I like to call gun tags.

A gun tag is nothing more than a nickname for the gun, instead of using its full name all the time. That just gets tedious. You may have noticed I've been using tags throughout this book. Let's look at an example in this context.

I have a Beretta Model 92fs chambered in 9mm. The writer who uses a generic semi-auto can only use:

- Pistol
- Handgun
- Gun
- Firearm
- Weapon

Since I've established what the gun is, I can add these options to my repertoire:

- The Beretta
- The Model 92
- The 9mm
- He pulled out his nine* (next page)

I probably wouldn't use *Model 92*; it just doesn't have good flow. However, with some other guns, the model number is a great option. If I had a SIG Sauer P226 or a Shadow Systems CR 920, it would look and sound better.

She pulled her Model 92.

He pulled his P226.

They pulled their CR 920.

Often, gun aficionados will drop the model designation in conversation.

- Beretta 92
- SIG 226 (Mr. Sauer's name is regularly dropped, poor guy.)
- Shadow Systems 920

I like to use the manufacturer's name more than anything, but some of them are a little wonky. I'd never use a Shadow Systems firearm in a story because the company name isn't easily recognized and could be confused out of the proper context.

Before I continue, I must circle around to the asterisked example from the previous page.

He pulled out his nine.

Many gun aficionados will shake their heads in disgust at this and say, "It's a 9mm, it's not a *NINE*."

To them I say, "Grow up and look at context."

It's all about context. If the character is involved in a gang or street life, they very well may call it a nine. However, purposeful use of slang should only be done in thought or dialogue, or if the story is a first-person narrative. Even then, you'll need to inform the reader of what the slang term means.

Tags work for rifles too. However, they're mostly restricted to the manufacturer's name or the model. I'd never refer to a rifle by its caliber. Let's look at the Bushmaster AR-15 chambered in 5.56mm.

She fired the Bushmaster.

They fired the AR.

He fired the 5.56.

Using the caliber tag just seems off; although, it might work for large caliber rifles or shotguns.

Gun tags and epithets can work for or against you. If your story has two different guns made by the same manufacturer, you could confuse the reader if you use the manufacturer as a tag.

He fired his SIG at the bad guy.

Reader: Did he mean the P226 or the 1911?

Remember the gun choice traps when using gun tags. It bears bringing up Glock again because I've seen reviews where model number/caliber confusion was an issue. When writing Glocks, establish the model number early, then never use it as a tag. Especially, if it's one of those that can be confused with caliber.

To learn which tags are used for specific guns, go to a gun store and talk to a salesperson.

Writer: What can you tell me about the SIG Sauer P226?

Salesperson: The two-two-six? SIGs are really popular. The state troopers are armed with SIGs. This baby is chambered in nine-mil ...

Using tags properly can give you instant credibility with your informed readers, even if they don't consciously register the usage.

Interlude: Ghost Guns

Guns that are privately made or assembled using unserialized parts are considered ghost guns. Every legally manufactured firearm has at least one part that holds that gun's serial number. In rifles, it's usually the receiver. For handguns, it's the frame.

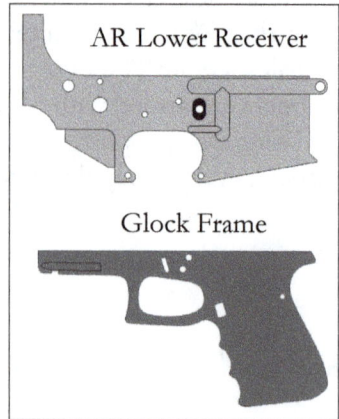

AR Lower Receiver

Glock Frame

In AR variant rifles, the lower receiver is the only part with a serial number.

Every other part or component either mounts to or is housed in the receiver or frame and can be purchased separately.

Ghost guns can be ordered off the internet as a kit and have no serial number.

If a person has the proper blueprints, receivers and frames can be manufactured using 3-D printers or a machine shop.

When a person purchases one of these ghost components or kits, they circumvent all firearms laws and regulations. There is no registration or background check.

Obviously, guns like this are a problem for law enforcement and they should be illegal. Zip guns fall into this category.

(Sorry, I'll not be providing details on how to construct a zip gun – you can foul up your own internet search history.)

Zip guns consist of a barrel, firing pin, and trigger mechanism. Most zip guns are constructed using unrifled pipe. Their accuracy sucks, but, then again, they are used at very close range.

The biggest concern when constructing any of these firearms is using materials in the chamber and barrel that are strong enough to handle the pressure of firing the projectile without breaking apart in the shooter's face. (See video link below.)

The zip gun will be used by low-tech ruffians, while the ghost gun will be used by the discerning criminal, villain, or spy.

Video:

Using Stronger Ammunition than the Gun Can Handle:
https://www.youtube.com/watch?v=gnVf62hOt-g

The Style Guide

I've sprinkled firearm advice throughout the informative sections of this guide. Here's where I'll collate it all and apply it in a manner similar to *Strunk & White's Elements of Style*. Some of this information will be repetitive to what came before. It can't be helped. I'd rather repeat information than have you hunt through the pages looking for that one nugget that will help you. I'll be providing links back to relevant sections to aid you if the advice needs more clarification.

For examples that show a side-by-side comparison, the least preferable option will be on the left. If that example has a mistake, the mistake will be bolded.

For single sentence examples, mistakes will be bolded, and an explanation for why it's wrong will follow.

Final Clarification

There will be places where I say, "Never/do not/don't do this." That advice is for your narration, not your characters. However, if your character is a professional, all the *nevers* and *don'ts* apply. If your character is a novice or inexperienced, they can get as much wrong as you want them to. Just be sure to let the reader know this is intentional.

Scenario: A firearms novice is getting advice from an expert. The firearm is a Glock 19.

Novice: *So, I put the clip in here?*

Expert: *It's a magazine, and yes that's where it goes.*

Rules of Usage
Know Your Gun's Characteristics

The minimum information you need to know about any firearm to include it in your story is manufacturer, model, and caliber. This will allow you to use gun tags and epithets. (Page 241) This gives your prose more variety rather than being shackled to using only common nouns such as gun and bullet.

Scenario: The police officer is armed with a Smith & Wesson .44-caliber revolver.

He fired the revolver.	*The magnum roared in his hand.*
He drew his gun.	*He drew the big Smith from its holster.*
The bullet punched a hole in the wall next to the villain's head.	*The .44-caliber round punched a hole in the wall next to the villain's head.*

Know the Gun's Anatomy
Gross Anatomy
Knowing your chosen firearm's anatomy is critical if you're going to add specific actions or procedures. The worst mistake you can make is an anatomy mistake, which usually occurs when the writer gives the gun a feature it doesn't have. You'll get called out on these mistakes 100% of the time.

Scenario: The character is armed with a Kimber K6s

The fight was over. She lowered the hammer and let out a sigh of relief.
Incorrect. The K6s is a hammerless double action only revolver. You must ensure that the firearm has the relevant part if the character is going to manipulate it.

Safeties
One of the most common anatomy traps a writer can fall into is mistaking a trigger safety for a safety that can be positively set. (Page 167) Most firearm aficionados use Glock as the primary example for this mistake. However, as more gun manufacturers adopt this feature, the pool of possible offenders gets larger.

Examples of pistols equipped with a trigger safety

- Glock 19
- Shadow System CR920
- H&K P30 Variant 1

Examples of pistols equipped with a positive safety

- M1911
- Beretta 92f
- H&K P30 Variant 3

Scenario: The character is armed with a pistol (Use of gun tags in the examples).

She flicked off the 920's safety.
Incorrect. The Shadow Systems CR920 has a trigger safety, which doesn't count as a safety in the gun world.

She flicked off the Beretta's safety.
Correct. The Beretta 92f has a positive safety.

She flicked off the Sig's safety.
Incorrect. The Sig Sauer P226 doesn't have a safety.

Nomenclature

Another anatomy trap lies in using the wrong nomenclature while describing a character's interaction with the gun. This is a common mistake because different manufacturers have different names for parts that perform the same action.

The part that retracts the bolt for an automatic rifle:
- AR variants – Charging Handle
- Steyr AUG – Cocking Slide
- MP5 – Cocking Lever
- Thompson SMG – Actuator
- M1 Garand – Operating Rod

The part that stops or releases the slide for a semi-automatic pistol:
- Beretta – Slide Release
- Glock – Slide Stop
- Shadow Systems – Slide Lock

Write around this problem by using the action the character is performing rather than naming the part.

Scenario: The character is armed with an MP5 submachine gun.

She pulled back on the **charging handle.** Incorrect: Nomenclature	*She retracted the bolt.* Correct: Action without nomenclature

Scenario: The character is armed with a Glock 19.

He thumbed the **slide release,** *chambering a round.* Incorrect: Nomenclature	*He released the slide, chambering a round.* Correct: Action without nomenclature

Magazines versus Clips

Repeating firearms are equipped with either an internal magazine or a magazine well, which will house the ejectable magazine when it's inserted. (Page 73)

Clips are devices used to hold ammunition in a manner that makes it easier and faster to load into a magazine.

Clips feed mags, mags feed guns.

Scenario: The soldier is armed with an M1 Garand.

"Hey, buddy, toss me a clip. I'm almost out of ammo."
Correct. The M1 Garand uses en bloc clips to load its magazine.

Scenario: The soldier is armed with an M16.

"Hey, buddy, toss me a clip. I'm almost out of ammo."
Incorrect. The M16 uses ejecting magazines, and the character is a soldier.

There is nuance here. The term clip is in the lexicon and untrained or inexperienced shooters will sometimes use it. However, professionals will never confuse the terms. Consider the character's background. In the song, "Get 'Em" by Lil Wayne, he sings about having "one pistol, two clips."

You may use the term clip instead of magazine in dialogue if it's consistent with the character's level of firearms training.

Never confuse the terms clip and magazine in narration.

Scenario: Scott and Chris are gangsters armed with Glock 19s (ejecting magazines).

Chris: "Scott, toss me another clip."
Correct. The sentence is dialogue.

Scott tossed Chris another clip.
Incorrect. The sentence is narration.

Know How the Firearm Functions

Many firearms mistakes occur simply because the writer is unfamiliar with the way a particular firearm operates.

Click on Empty (Automatic rifles, page 86, semi-auto pistols, page 159)
The gun clicked on empty, is one of those universal phrases that is wrong almost all the time when applied to semi-auto firearms.

If the slide or bolt locks to the rear, the gun will not click on empty.
Guns that don't click when empty:
- Almost all semi-automatic pistols
- Almost all semi-automatic/automatic rifles that fire from a closed bolt
- Guns that can click when empty:
- Almost all manually operated rifles
- All revolvers

Guns that *thunk* on empty
- Almost all firearms that fire from the open bolt

There can be no definitive rule for guns clicking on empty because there are always exceptions. However, there are guidelines.
- If the semi-automatic firearm's bolt or slide locks to the rear on the final round, the gun will not click when empty.
- When the action of a firearm is open, except those that fire from the open bolt, the trigger is disconnected. The gun will not fire, the gun will not click.
- Guns that fire from the open bolt will go *thunk* when they're empty if fired in the fully automatic mode.

Menacing
Menacing by performing an action with a firearm is regularly ridiculed in the firearms community and should be avoided.

Scenario: The hero is cornered by the villain after a firefight. The villain is armed with a pump-action shotgun.

"You'll talk." He racked the shotgun. "Or I'll blow off your foot."
Incorrect. One of two things will happen when the villain racks the shotgun:
- A shell ejects – The gun was already loaded, and the villain looks like an idiot.
- A shell doesn't eject – The gun was or is empty, and the villain looks like an idiot.

Scenario: The hero is captured by the villain, who is armed with a Beretta 92f.

"You'll talk." The villain cocked the hammer. "Or I'll blow off your foot."
Correct with caveat. Sure, the villain can perform this action. However, any double-action pistol is ready to fire as long as the safety is off. Cocking the hammer serves no purpose.

Likewise, if armed with a single-action pistol like the M1911, the hammer would have been in the cocked position already, if the gun was loaded. If the hammer was down on a loaded M1911, firearm aficionados would lose their minds since this is not the recommended carrying configuration for this gun.

Know How the Firearm Is Staged (Page 113)
This is subjective and based on the character's preferences or the SOPs of the character's employer. A firearm is considered to be staged if the character is not carrying it (e.g., under the bed, in the glove box of a car, etc.). A staged firearm is usually in some state of readiness – loaded or cruiser-ready.

Know How the Firearm Is Carried
Rifles, shotguns, and machine guns are easy. When going into action, the firearm will be loaded and on safe until ready to shoot. Caveat special purpose firearms. (Page 114)

Handguns are more difficult. Manufacturers give guidance to how their firearm should be carried, and agencies have policies that guide how their members should carry their firearms. The character will have their own preferences for how they carry their firearm. (See page 177 for more information.)

Know Your Terminology
Longarms
Longarms are firearms designed to be held in both hands and fired while braced against the shoulder. While shotguns and rifles are both longarms, they are not the same type of firearm. Don't confuse the two.

Carbines are shorter versions of a rifle. The M4 is a carbine version of the M16.

Scenario: A soldier inspects her M4 carbine.

She picked up the M4. The new rifle was lighter and shorter than her old M16a2. Correct with caveat. While a carbine is by definition a rifle, do not treat the terms as interchangeable. (Page 94)

Handguns

Use terminology that is both correct and won't confuse the reader.

Scenario: The character is armed with a .357 revolver.

She tucked the pistol in her belt.
Correct with caveat. While a revolver is considered a pistol by definition, the writer should never use pistol when writing about a revolver.
- Use pistol for all semi-auto handguns.
- Use derringer for any handgun designated as such.

Suppressors versus Silencers (Page 200)

The correct term is suppressor.

Scenario: The inexperienced witness is talking to the police.

Witness: *"The assassin had a silencer on the end of his gun."*
Correct. As with clips and magazines above, characters who are inexperienced or ignorant of firearms and correct terminology are allowed to make mistakes in dialogue.

Use the proper term in narration.

Rules for Ammunition (Page 184)

Ammunition can be expressed in either the metric or imperial measurement systems.

Metric Measure

Metric ammunition is expressed in millimeters and never has caliber attached.

Scenario: Two police officers are talking about the caliber of a gun used in a murder.

"The weapon was 9mm caliber." Incorrect.

"The caliber of the weapon was 9mm." Correct.

Never use a period in front of metric ammunition. If you write .9mm, you're expressing the measurement 900 microns – which is the size of a fiber optic tube.

Rifle ammunition, when expressed numerically, always retains its decimal marks 7.62x39mm.

Imperial Measure

Imperial ammunition is expressed in hundredths of an inch and may have caliber attached.

There is no standardized way to write caliber. All the examples below are technically correct.

.22 caliber   22 caliber

.22-caliber   22-caliber

.22

When writing narration, I prefer .22-caliber. Readers will recognize the ".##" as a caliber expression, and the hyphen keeps "caliber" from dangling out in space.

The gun was a .22-caliber.

22-caliber is also acceptable given narration issues described below.

The format you decide to use is not as important as remaining consistent and differentiating between narration and thought/dialogue.

Appellations (Page 186)

Caliber appellations are important for identifying whether ammunition is compatible with a certain firearm. Appellations are always capitalized. For .22 Magnum, *Magnum* is the appellation. And for .22 LR, the appellation *LR* stands in for *Long Rifle*. A .22 Magnum cartridge will not fit in a gun designed to fire .22 LR.

It gets confusing when the appellation is the name of a firearms company: .223 Remington is a caliber that can be manufactured by any ammunition company… not just Remington.

Writing ammunition caliber in thought or speech can be tricky. Secondary information about the cartridge is often dropped. Especially for common calibers. Thus:

- 9x19mm Parabellum becomes 9mm or *nine-millimeter.*
- .30-06 Springfield becomes .30-06 or *thirty-ought-six.*
- .45 ACP becomes .45 or *forty-five.*

In context, .45 is understood, by firearm aficionados, to be pronounced *forty-five.* However, some people may read this as *point-four-five.* This is especially true of metric rifle ammunition. Professionals will think or pronounce 5.56 as *five-five-six.*

For the average reader, how ammunition is written in dialogue isn't important. However, this distinction is important for an audiobook version. An intense scene can be ruined by a decimal point.

Scenario: The soldier is in a firefight. He needs more ammunition.

"Hey, Sarge, throw me another mag of 5.56."

The written passage is correct. However, the narration may sound like this:

"Hey, Sarge, throw me another mag of five-point-five-six."

This sentence can be fixed by either removing the caliber:

"Hey, Sarge, throw me another mag."

Or writing the caliber as you wish it to be spoken:

"Hey, Sarge, throw me another mag of five-five-six."

Caliber Slang

Sometimes caliber is shortened to cal in dialogue. While it can be used universally, I find it most natural with two-syllable calibers. Three-syllable calibers tend to get the full word caliber.

When talking about a .40-caliber, I find it's more natural to say, *"Forty-cal."*

When talking about a .45-caliber, I find it's more natural to say, *"forty-five,"* or *"forty-five-caliber."*

Weird calibers like .380 ACP tend to lose caliber altogether. I'd say, *"three-eighty."* Never *"three-eighty-cal,"* or *"three-eighty caliber."*

Shotgun Ammunition (Page 203)

Shotgun ammunition is expressed gauge. As with rifle and handgun ammunition, there is no industry standard for writing the size. It can be expressed as either 12 gauge or 12-gauge. As above, I prefer using the hyphenated version.

- Never add caliber after shell size.
- Never add a decimal in front of gauge. The exception to this rule is the only shotgun sized in caliber – the .410 pronounced four-ten.
- The .410 always gets its decimal when written in number form.
- The .410 is never followed by gauge or caliber. (.410 is just .410)

Shotgun Appellations (Page 206)

Shotgun appellations express the size and amount of shot contained in the shell. With shotgun ammunition, once the gauge of the shotgun is established, it's the gauge that's often dropped instead of the appellation.

Question: *What ya shootin?*

Answer: *Double-ought buck.*

The information that is important to the conversation is the size of the shot.

Scenario: A police officer talking to the SWAT team's breacher. Breaching gun is a Mossberg 500 (12-gauge). (Shotgun Breach, page 116)

Officer: *What type of ammo do you use?*

Breacher: *Number-six-bird.*

Use Generic Firearms
Only firearms that are important to story, plot, or character need to be identified. If the gun only shows up once and is never seen again, don't invest a lot of effort into it. Give it the generic noun and move on:

The character killed himself in the prologue with a revolver.

He reached down and picked up the robber's gun, a cheap pistol.

Remember that caliber bias is a real thing in the gun world. Smaller caliber guns are often dismissed as being cheap and of poor quality. Therefore, the second example from above could be written:

He reached down and picked up the robber's gun, a piece of crap .32.

Use Common Firearms
If possible, use only recognizable brands. Uncommon firearms and brand names always attract the attention of the gun aficionado and are more likely to be investigated.

The hero drew their Grendel from the holster.

The hero drew their Glock from the holster.

Common firearms are also easier to research.

That Pesky Hyphen
Part One:
Manufacturers are not consistent in their use of a hyphen when assigning model numbers to their firearms. Be sure to double check the proper form of the firearm's model number. Adding or omitting the hyphen can make or break your credibility in the firearms world.

Part Two:
Style rules, dictionary definitions, and common usage often contradict when it comes to naming firearms. Quite like the way caliber is expressed, there is no industry standard. When talking about a certain shotgun, both the open compound *pump action* and the hyphenated compound *pump-action* are correct.

My preference is to use the hyphenated compound when it's modifying a noun: *pump-action shotgun*.

Or, if the compound is used as a noun: *Semi-autos* are great pistols.

Finally, use the hyphen when the noun is implied:

Seth: What kind of rifle was it?

Bart: Some kind of lever-action.

In this instance, lever-action describes the unnamed rifle. Of course, the other formation can be equally correct.

Seth: What kind of rifle was it?

Bart: Some kind of lever action.

In this instance, Bart could just as well be describing the action rather than the gun as a whole, which would mean the hyphen should be dropped.

I'm giving you my preference. You are free to use whichever form you prefer, but I caution you to remain consistent throughout a single body of work.

Principals of Composition
Keep It Simple, Keep It Safe
Generic actions should be used whenever possible.

The character *loaded/fired/reloaded/aimed/unloaded* their firearm.

This will reduce the likelihood of making either an anatomical, procedural, or nomenclature mistake.

Training Scene

If you're going to demonstrate a character's firearms knowledge and expertise, you should only do this in a training scene. These scenes are heavy on process and nomenclature, which takes away from the emotional impact of an action scene.

Scenario: The character is at the gun range. She is armed with a Glock 19.

She locked the slide to the rear and inspected the chamber. Next, she inserted the magazine, making sure that it seated properly. Finally, she released the slide, chambering a round.

Correct. Every step in this process is accurate. Note that there's no use of nomenclature in the passage. Even when describing process, keep things as simple as possible.

Action Scene

Strip as much process as possible from action scenes unless you are using the process to convey emotional state.

Scenario: The hero is in a gunfight, armed with a Glock 19.

Slide lock – the gun was empty. He changed magazines and got back in the fight.

Or

Slide lock – the gun was empty. His hands shook as he tried to insert the new mag. Finally, when the gun was ready, he got back in the fight.

This technique of minimal firearm manipulation can be applied across all genres and firearm types.

Scenario: A cowboy is armed with a Spencer lever-action rifle. He's holding off bandits.

Seth fired and the bandit fell. Another bandit charged. He fired again. Two down.

There is nothing wrong with the above passage. The reader doesn't need to see every action your character makes. However, by adding just a little process, we have a better passage.

Seth fired and the bandit fell. He franticly worked the lever as another charged. He fired again. Two down.

By adding how Seth works his rifle, we can gauge his emotional state.

Seth fired and the bandit fell. He calmly worked the lever as another charged. He fired again. Two down.

Jams (Page 220)

Jam is in the lexicon; feel free to use it. Jam is a convenient way to say malfunction, and most malfunctions are easily remedied.

However, The gun jammed is an overused trope. Look for inventive ways to take the gun out of play. You can do this by running out of ammo, losing the gun to calamity, or by damaging the weapon.

Choreograph Your Action Scene (Page 271)

The sequence of events you see in your mind may not be conveyed to the page. Create a simple diagram. This will remind you where the character is when they fire rounds – helping with round count, reloads, or suffering injury or damage to their firearm.

Words and Expressions to Incorporate
Clear versus Unload

Most firearms professionals use the word *clear* as an action or a condition to replace unload and unloaded. When you use *clear,* you're speaking the language of the firearm aficionado.

He cleared the shotgun. – unload.

Is that weapon clear? – unloaded.

Dry versus Empty

Here we have another interchangeable set of terms. However, this one is a little more nuanced.

Empty is more appropriate in narration:

Their gun was empty.

Dry is more appropriate in thought or dialogue:

"Hell, I just fired the AR till it went dry, then grabbed my pistol."

Slang Terms

Feel free to use any slang term you want. However, these terms should only be used in dialogue.

Character: *"I pulled out my nine and busted a cap in his ass."*
Correct. Dialogue.

Chris pulled out his nine and busted a cap in his ass.
Incorrect as narration. The previous sentence should be:

Chris pulled out his Beretta and shot the man.

Firearms, Actions, and Expressions to Avoid
Confusing Guns (Page 229)

Glock model numbers can be confused with caliber. If you write Glock 22, do you mean the .40-caliber? Or do you mean the Glock 44, which is the .22-caliber version? Keep your terminology clear and use a caliber format that removes confusion.

Remember that some guns have variants of the same model. The H&K P30 comes in variants 1 and 3. These variants have vastly different features.

Cordite

Scenario: The SWAT team has just raided a terrorist hideout in modern-day New York.

The firefight was over. Spent brass littered the ground and the stench of cordite filled the air.

Incorrect. Cordite was a propellant used in the manufacture of gunpowder from 1889 to 1941. You may use the word cordite to describe the smell of gun smoke only if your story falls within this period. Otherwise, use gunpowder or gun smoke.

Getting Overly Technical

As you become more familiar with firearms, you'll run into terminology and lingo that you may be tempted to use. Don't. While you want to appeal to your firearm aficionado readers, some juice just isn't worth the squeeze.

Battery is a term that describes when the bolt or slide is fully forward and locked in place. If the gun is in battery, it's ready to fire.

I use the term only twice in this book. Here, and in the glossary. Taking time to explain a term to the uneducated segment of your readers kills the flow of your story. Why explain battery when, "the gun was loaded" will suffice?

Going Hot/Go Hot/Weapons Hot (Page 112)

This phrase is used to tell soldiers or police officers to take their longarms off safe before a raid or action. Most modern operators keep their weapons on safe until they need to fire. After the engagement is over, they immediately put the firearm back on safe.

Diving and Rolling

Diving is acceptable if diving behind cover, but diving to achieve the optimal firing solution is a hard no.

Effectively engaging a target while walking is hard. Engaging a target while diving is nearly impossible. Then, of course, you hit the ground with your arms extended. A great way to knock the wind out of yourself.

When I say rolling, I mean somersaulting. Rolling left or right, when prone, to avoid gunfire. This is valid. Everything else is a hard no for any genre.

Look at it this way, if I'm walking or crouching while moving from one place to the next, I can see and make adjustments to my course.

If I'm diving, I've committed, and I'm locked into the outcome. If I roll forward, I lose sight of the world around me and can be momentarily disoriented when I return to a crouching position. There should be no gymnastics in a gunfight.

Effects of Firearms (Page 187)

Bullets penetrate the body, tear muscle, rupture organs, and break bones. They cannot hurl someone across the room.

Scenario: The character is armed with a sawed-off shotgun; loaded with 12-gauge slugs. He is engaging his nemesis.

He fired from a distance of five feet. The slug punched into the villain's chest, throwing him backward into the wall.

Incorrect. It's physically impossible for a bullet to impart enough energy to the human body to throw it across a room. Write instead about wounds and their effects on the character.

Two Pistols

No, no, no. Lara Croft can use two guns. Antonio Banderas can, too. You may not.

I've been in a lot of Simunition firefights against all manner of people. Every now and then, I would try something out just for fun. One thing I tried was fighting with two pistols.

The first thing I found out: You can't aim both guns at the same time, even when directly to the front. The gun in my dominant hand was accurate, but I had no idea where the other one was shooting. I ended up engaging my targets with either the left-hand pistol or the right, depending on angle and cover. That worked great until I had to reload.

That was the second thing I found out. It's really hard to reload a pistol if you have a pistol in your other hand.

Using two pistols depends on genre. If your work is an over-the-top fantasy, go for it. You've probably written in cool reloading procedures and devices to make it work. If you're going for realism, don't.

Political Terms

Political terminology is often composed to elicit an emotional response in regard to firearms. These terms are often ill-defined, or their definition varies from one legal jurisdiction to the next. Firearms aficionados use clearly defined *industry* terms when talking about guns.

Your novice characters can, mistakenly, use some of these terms. While your politically active character may use them exclusively. Here are a few examples:

- **Weapon of war.** Anything can be a weapon of war. My grandfather's Ka-Bar was a weapon of war. My entrenching tool could be used as a weapon. Spears, bows, crossbows, and swords were all weapons of war at one point.
- **Assault weapon.** Anything can be a weapon to commit assault. Professor Plum would tell you that the candlestick in the conservatory makes an excellent weapon for committing assault/murder.
- **Assault rifle.** See assault weapon above. Also, do not confuse AR – Armalite Rifle – for assault rifle. The firearms enthusiasts will write scathing reviews in response to this error.
- **Cop killer bullet.** Any bullet is a potential cop killer. (Page 199)
- **Military-grade.** Which military, what era? You do know that gear was made by the lowest bidder?

I understand the current climate and what the terms mentioned above are meant to represent. However, there is a segment of the population that will make the same arguments as above.

The 99% Solution

This technique avoids all possible mistakes in writing about the gun or its use.

If you name the gun in your story, use gun tags (Page 241), and keep your character interactions limited to load, reload, aim, fire, unload, and holster (pistols), you'll not invoke the ire of firearm aficionados.

As you add more detail to the gun, or the character's interaction with it, the risk of making a mistake goes up. Additionally, mentioning certain gun-handling techniques will attract the attention of the experienced shooter.

I call this the 99% solution because there's always someone out there that will find fault with even the vaguest of actions. This is because the firearms community is full of biases.

Biases

There are four main bias categories in the firearms community. The main way to avoid these biases is to talk to a professional. However, talk to four professionals, and you'll get five opinions.

Manufacturer Bias

If you go on cheaperthandirt.com, a shooting sports discount site, then click handguns by brand, you'll find over one hundred manufacturers listed. The best way to avoid a manufacturer bias is to stick to well-known brands. Which would you trust more, a handgun made by Colt, or one produced by Zastava Arms USA?

I don't know what Zastava makes. I didn't click on the link. Nor have I viewed any reviews of their products, but I know what my choice would be.

Equipment Bias

There are a million different holsters, optics, sights, and modifications out there for every firearm. Some are better than others, but at some point, it boils down to personal preference.

Remember, a holster is just a holster, and a scope is just a scope, until you name it.

Technique Bias

This bias runs the gamut from how to properly manipulate the firearm to how it's carried.

Scenario: The hero is in a gunfight. He's armed with a Glock 19.

Slide lock – the gun was empty. He dropped the empty magazine, slammed in a new one, thumbed the slide stop – chambering a round – and got back in the fight.

Correct. This sentence is correct in all ways. However, it bumps up against a technique bias. (Page 181) I could write the sentence using the other technique, but I'd be just as wrong with the other half of the community. The only way to please both is to be vague:

Slide lock – the gun was empty. He dropped the empty magazine, slammed in a new one, released the slide – chambering a round – and got back in the fight.

I don't tell the reader how the character released the slide, only that the action was performed. I let their bias fill in the blanks for me. Let's look at a carry bias.

Scenario: The cop is armed with a Beretta 92 (Double to single action: DA/SA)

She de-cocked the pistol, flicked it back to fire, and holstered it.

Correct. There is nothing wrong with the character's actions in the above sentence. This is the proper operation of the Beretta. However, there are some that believe that if the pistol has a safety, it should be used. (Page 179)

When looking at carrying conditions, page 177, we see that the DA/SA handgun should be de-cocked before holstering. We can remove the position of the safety lever and be good to go:

She de-cocked the pistol and holstered it.

This is better. By saying she de-cocked the pistol, the writer lets the reader know that they know the pistol is double action and needs to be de-cocked. The rest is left to the reader's imagination. Or ...

She made the gun safe and holstered it.

Even better. The writer acknowledges that the DA/SA gun needs some attention before it can be holstered. What that action is, is left for the reader to decide. And then of course there's this ...

She holstered her pistol.

That last one is my favorite. Readers will fill in the blanks based on how you've built up this character's training and experience. You can cement this in by adding the training.

She holstered her pistol like she was taught in the academy.

Now we've sidestepped all biases. Referring back to training versus action scenes. The sentence above is great for a training scene. The sentence below is more fitting for an action scene.

She holstered her pistol.

Caliber Bias

The .32-caliber round is widely considered to be an inferior caliber. Thus, all guns chambered in that caliber are considered inferior. Pistols that fire that round are, also, comparatively inexpensive – another strike against.

A caliber snob considers all calibers that aren't their favored to be inferior. These folks usually revere .45-caliber.

"If it don't start with a four, it ain't worth a shit."

You can see my thoughts on the matter on page 189. Personally, I don't want to get shot by any caliber.

Final Advice

1. Handle or fire the gun if possible. Nothing beats experience. In places with strict gun laws, this may be difficult. In some places, you can rent guns to shoot on a range. This is the optimal solution. Again, your circumstance will vary based on your location.

2. Talk to a gun expert. Preferably one who has experience carrying a firearm for a living. People who work in gun shops are a good source of information on function, but they may not be the best source for application.

3. Cultivate a source in an agency similar to the one you're writing about. Most people will be happy to help for nothing more than a thank you in the acknowledgments. Be careful here to make sure they are era-appropriate. Techniques have changed a lot in the last thirty years.

4. Read accounts of real-life firefights.

5. Use movies and TV shows for inspiration, not information.

6. When in the groove, don't sweat the details. Use your preferred text marker and keep on rolling. You can fill in the proper details later.

7. Have fun.

Writing the Gunfight

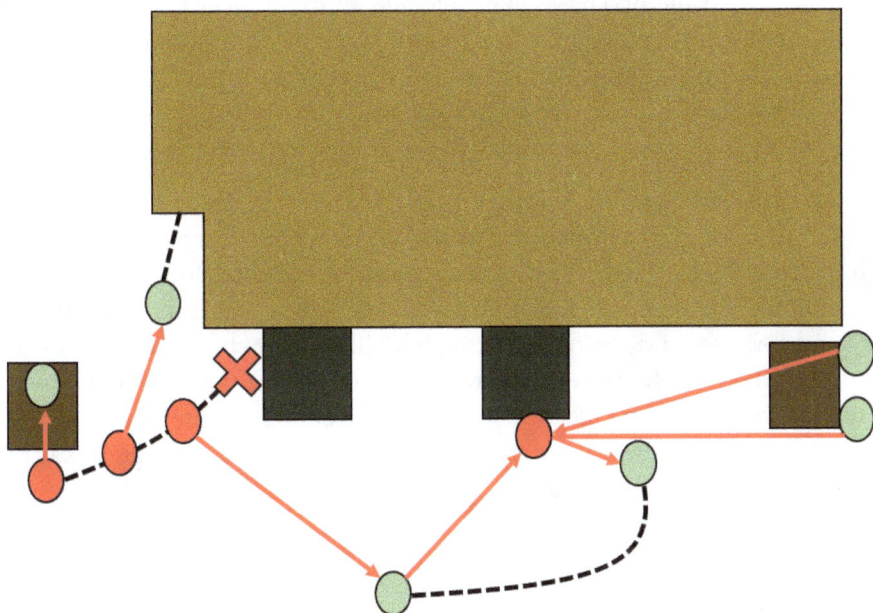

If you've read this book from the beginning to this point, you've got a good idea of how firearms work and what their effective ranges are. You may have even selected a firearm or two to use in your current work. Now we need to bring it all home and plan your mayhem.

Spatial Considerations

The FBI and several law enforcement agencies publish data every year that captures law enforcement officer-involved shootings. Apart from that, there's a Rule of Threes. The average gunfight takes place under three yards, combatants fire three rounds, and the fight lasts three seconds. However, general consensus holds that most gunfights take place at distances from three to five yards.

Let's look at a few case studies to put things in perspective.

Truman Assassination Attempt

In 1950, two men attempted to assassinate President Truman at Blair House, where the president was in residence while the White House was undergoing renovations. This fight involved eight combatants, lasted around forty seconds, and approximately thirty-four shots were fired – a little less than one shot per second (averaged). When the fight was over, one officer and one would-be assassin had been killed. The other assassin and two officers were wounded. The longest engagement distance was about twenty feet (seven yards).

The Miami Shootout

In 1986, eight FBI agents attempted to arrest two bank robbers. The fight involved ten total combatants, lasted about five minutes, and approximately 145 shots were fired – about two shots per second (averaged). By the end, both bank robbers had been killed, two agents had been killed, and five agents had been wounded. The longest engagement distance was about thirty-five feet (twelve yards). One firearm was rendered inoperable by bullet strike.

The North Hollywood Shootout

In 1997, hundreds of Los Angeles police officers became involved in a gun battle with two bank robbers. The fight lasted forty-four minutes and more than 1,600 rounds were fired – almost two shots per second (averaged). At the conclusion of this fight, both robbers had been killed, twelve officers and eight civilians had been wounded. The longest engagement distance was over 350 feet. One firearm was rendered inoperable by a bullet strike.

Let's look at what's important to the elements of a gunfight ... as far as you, the writer, are concerned.

We have the **Rule of Threes** that states one shot per second. This bears out if you average the round count of our three case studies. However, long duration fights comprise periods of time where one party or the other is assessing the situation and moving to a position of advantage. These actions culminate in a flurry of shots that decide the outcome of the battle.

Distance is important, but not critical unless your character is pushing the limits of their firearm (page 208). People move in a gunfight, either to get an advantage on their opponent, to seek cover from them, or run away. They crouch, dodge, retreat, and advance based on their immediate assessment of the situation. This drives up the round count of a fight because it's hard to hit a moving target. It's even harder when you're moving too.

The duration of the fight determines how much time a person has to think. Three feet, three shots, three seconds equals no time to think. At that distance with short duration, training takes over and actions are instinctive. Extended time and distance allow for thought, analysis, and tactics.

Other Factors

Physiologically, there's a lot going on in a person when there's sudden violence. The body has what's called a sympathetic nervous system response, fight or flight. In addition to rapid heart rate, the person can experience:

- Perceptual narrowing – tunnel vision
- Auditory exclusion – diminishment of the ability to hear
- Time dilatation – feeling like things are happening in slow motion
- Diminishment of fine motor skills

Before I go any further, I've got to warn you that if you research the SNS response in gunfights, you'll run into a lot of quasi-science. Yes, the things mentioned above happen. However, in my experience, training level determines to what degree.

A normal person doesn't get in many fistfights. Thus, when they suddenly find themselves in one, they suffer a devastating SNS response. On the other hand, a boxer gets in a fistfight every day they spar. For them it's just Tuesday.

The same holds true for gunfights. Trainers use paint rounds that allow soldiers and officers to experience the gunfight environment without lasting injury. I used to call these exercises sparring.

So, your character may experience the effects of an SNS response, but it may not be as devastating as some people say.

Fight Scene Guidance:

- Fast fight/no thought. Write this type of fight with quick brutality. The faster the fight, the closer the distance, the less the thought. Characters don't have time to think about anything other than survival. Deal with any emotional issues afterward. (See Truman assassination attempt, page 268.)
- Medium fight/some thought. Limit these thoughts to the tactical situation. However, there's still no time to deal with emotions. (See Miami Shootout, page 268.)
- Long fight/a lot of thought. In long duration fights, there are natural lulls punctuated with moments of extreme violence. It's in these lulls you can work in emotional content. (See North Hollywood Shootout, page 268.)
- Use one shot per second as a guideline. (Page 269)
- People shoot multiple times in a fight. Keep a round count and don't forget to reload. (Reloading occurred in every example above.)
- Move your characters during the fight. They need to duck, dodge, and move when rounds impact close to them.
- Cover is anything that can stop a bullet – a bush cannot. You can't say, They took cover behind some shrubs.
- Injure your characters. Injuries may be:
- Superficial: no impact on character capability
- Serious: possible impact on character capability
- Lethal: sometimes immediate death. However, some characters can function for a limited time after receiving fatal wounds. This happened in both the Miami Shootout (Page 190) and the Truman assassination attempt. (Page 272)
- Render firearms inoperable by bullet strike. This happened in both the Miami and North Hollywood shootouts.
- Use reference of the effects of an SNS response (Page 269) on your character sparingly. If they have tunnel vision, we can't see through their eyes.
- Choreograph the fight before you write it.

Choreography

I've been in thousands of simulated gunfights where the pain was real, but the cost wasn't permanent. In most of them, it was my job to provide an assessment of my opponent's actions. What they did right, what they did wrong, and why I made my decision to shoot at them. All while being shot at myself.

I knew the team's tactics. I knew when they were coming, and I designed scenarios to test group and individual discipline.

I often had role players and I'd tell them, "Stand here, you're looking for this. If it happens, or doesn't, shoot."

I choreographed the scene.

Choreography also works well for composing fight scenes in fiction. It may take a little time on the front end, but it will save you time in editing.

Basing on a Real Event

For my first example, I'll use the attempted assassination of Truman as the base for an intermediate action scene in a story.

Assassination Attempt

There were nine combatants in the incident:

Good guys:
- SSA Floyd Boring – Wounded
- SSA Vincent Mroz
- Officer Donald Birdzell – Wounded
- Officer Leslie Coffelt – Killed
- Officer Joseph Davidson
- Officer Joe Downs – Wounded

Bad Guys:
- Oscar Collazo – Wounded
- Griselio Torresola – Killed

The Fight

The plan was for Collazo to approach from the east and engage the officers there, then move into Blair House. Torresola would approach from the west, take out the guards at that location, then follow Collazo into the building. While the whole incident is considered one gunfight, it took place as two separate actions, and it's easier to follow events that way.

- Collazo walked past the guard shack, passed Birdzell on the street, turned, and attempted to shoot him in the back. There was some manner of malfunction with the gun, and Collazo ended up shooting Birdzell in the knee.
- Birdzell – lacking cover – limped out into the street. He drew his revolver and fired several times at Collazo. Davidson and Boring, once they realized what was happening, immediately drew and fired on Collazo. Then they sought cover behind the guard shack, where they continued to shoot at him.
- Collazo traded shots with the three men until his Walther ran empty. He performed a reload, then fell to the ground, ending this part of the fight. He'd sustained wounds from several ricochets, but never took a direct hit.
- As the battle on the east side of the building began, Torresola shot Officer Coffelt three times on his way past the western guard shack. He then saw Officer Downs – who was returning with groceries –, and shot him three times.
- Officer Downs was hit in the hip, back, and neck. Despite his wounds, he managed to make his way into a side entry of Blair House, which he barred behind him.
- After engaging Downs, Torresola noticed Birdzell shooting at Collazo from the street. He fired once, striking Birdzell in his good knee, taking him out of the fight. At this point, Torresola's gun was empty, and he stopped to reload.
- Despite being mortally wounded, Officer Coffelt exited the guard shack, braced himself against the wall, and fired one round that struck Torresola in the head – killing him instantly.

How would you write these events? Would you use one or multiple POVs? Complex fight scenes with multiple POVs are a pain in the ass both for the writer and the reader. They require clean paragraph breaks that take away from the action.

Block the Scene

To figure out the best POV, I block out the actions of each man. To do that, I need to know the terrain. This diagram of Blair House is not to scale and has been overly simplified for clarity.

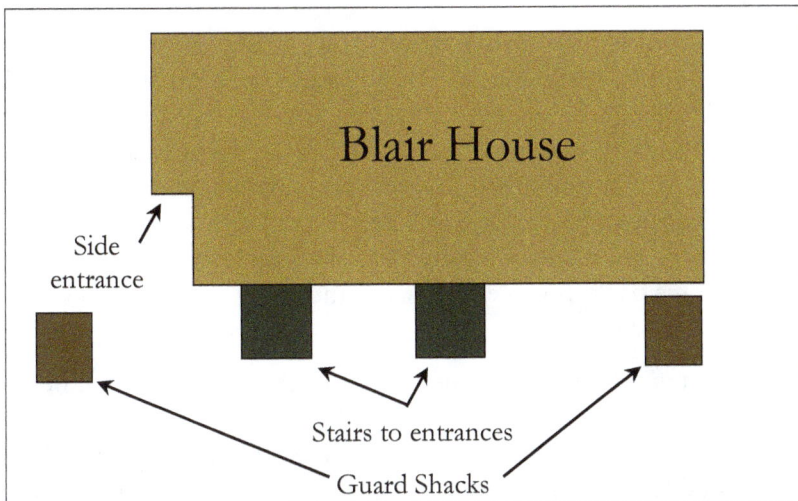

Next, I add my characters and their actions. Bad guys are red dots, good guys are green. If a character has multiple positions, they're connected by a dashed line.

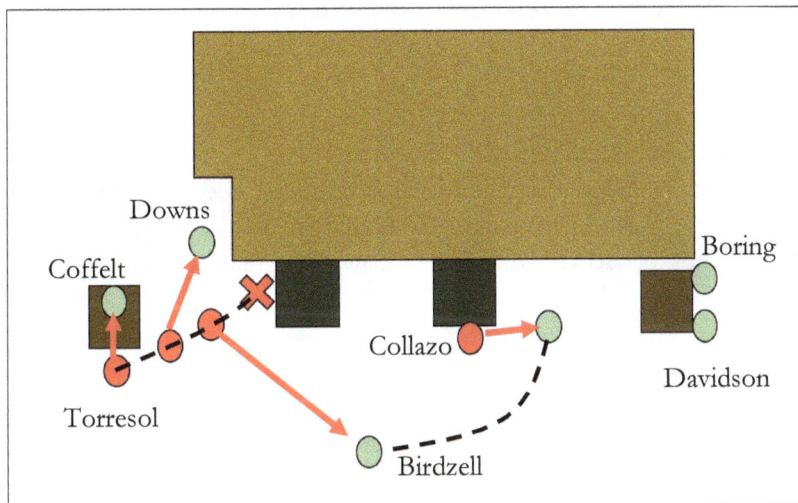

Clearly, Birdzell is the only good guy who has the ability to see the entire fight. Of course, in reality, he didn't. However, since we're using this as a base for our action scene, he does.

Simplify the Scene

I can now simplify my chain of events.

- Birdzell sees the man walk past him. He hears an audible click and turns to investigate. The man has a gun! Then, bang, he's shot in the knee.
- Move, need to move. He limps away as he draws his pistol. There's firing, a lot of firing. He sees Boring and Davidson shooting at the man. He shoots at the man. The man is down.
- Bang! He's hit in the other knee. He falls to the ground and rolls to his left. Another gunman! The second gunman is reloading.
- Movement catches his eye, and he sees a bloody Coffelt leaning against the guard shack. Coffelt fires and the other gunman falls. He rolls to his back. It's over, he sighs, and that's when the pain hits.
- Birdzell passes out.

Now I can write the scene. The actions of the other men can be filled in later.

That's how I do it if I'm basing my action scene on a real event. But what if I need to make my own? Don't worry. I've got a six-step plan to help you out. Steps one and two are fairly interchangeable.

Setting Your Own Scene

Step 1: Identify Characters and Equipment

- Officer Smith – M1911 – Hero
- Officer Jones – M1911
- Officer Tate – M1911
- Henchmen – bad guys (BG) Number TBD
- El Gato – Drug Leader (mid-level) target of raid – Glock 17

I've left the number of henchmen blank for now, because I haven't set up the chessboard yet, and don't know how many I may need.

Step 2: Identify the Objective of the Scene

If you have a fight scene in your story, it should serve either plot or character development. Random fight scenes with no purpose are just filler. However, using a seemingly random fight to achieve foreshadowing is awesome.

Purpose: Foreshadowing. Officer Jones is the true bad guy. Jones Kills El Gato and Officer Tate because they are liabilities to his scheme.

Step 3: Set the Chessboard

Knowing a floor plan, or a setting, will allow you to reasonably place characters where you want them and direct the flow of the action.

This diagram is a generic representation of a house I used to live in. Once I have the terrain, I can add in my pieces.

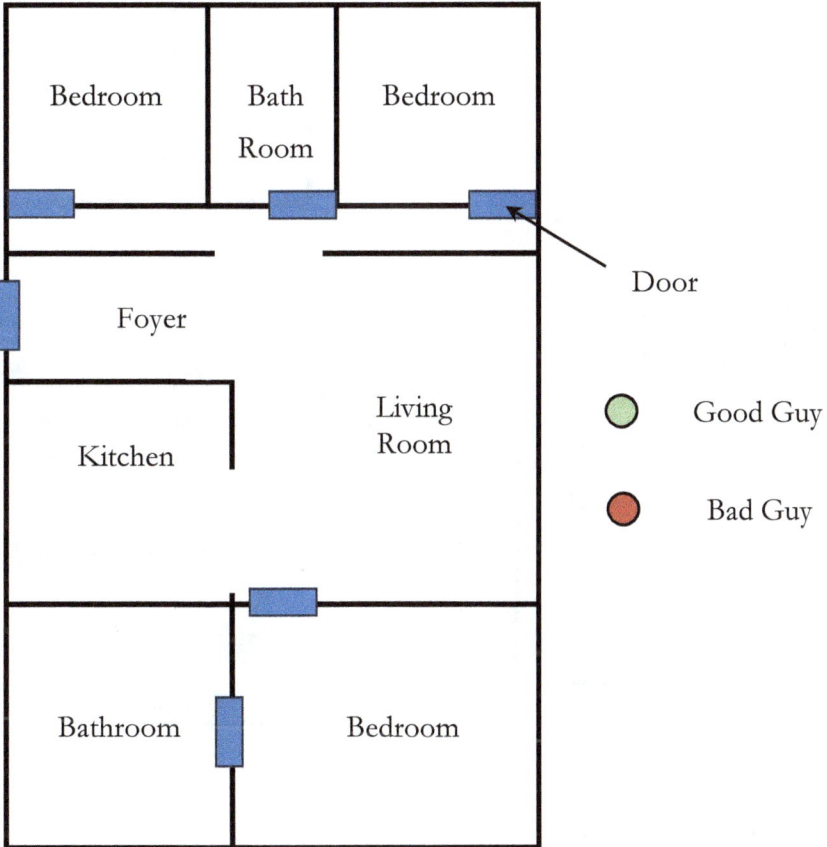

Step 4: Determine POV

POV: Officer Smith

Step 5: Plot the Action

Determine initial positions of the characters and plot the scene. Adjust as necessary.

- Smith, Jones, and Tate kick in the door.

- Smith shoots BG #1 in hall.

- Tate shoots BG #2 in living room

- Jones heads to the bedroom where El Gato is.

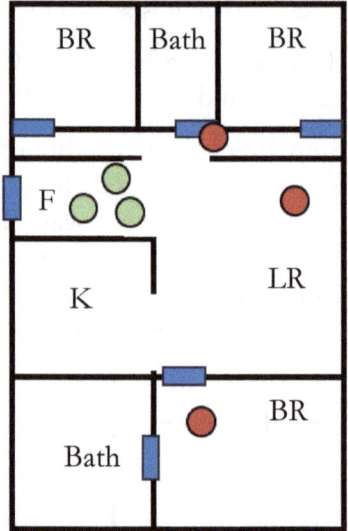

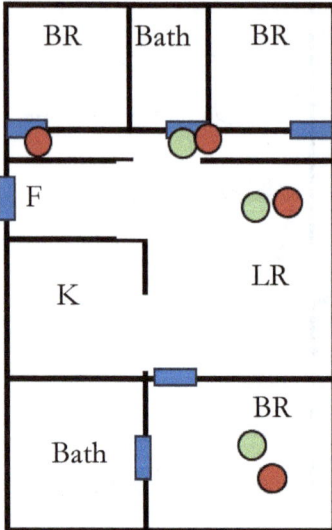

- Smith, shoots BG #3 coming out of BR. *Decision to add another BG to the scene. This keeps Smith in the hall longer.* He needs to reload.

- Tate yells at Jones for not clearing the kitchen.

- Jones is out of sight. Kills El Gato.

276

- Tate moves to clear kitchen.

- Jones kills Tate with El Gato's gun. He has line of sight on Tate through the open door. Smith can't see who fired the shot.

- Smith is left wondering what happened.

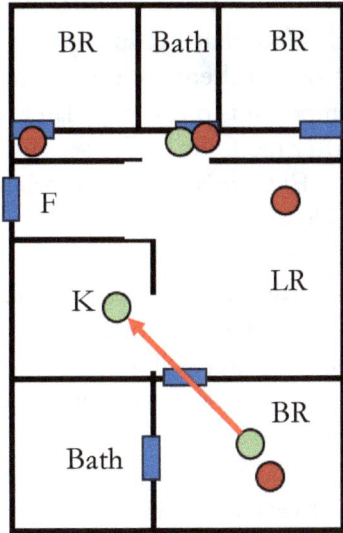

Step 6: Write the Scene

The example is a very simple scenario, and there's no need to kill any of the weapons in the fight. As an intermediate scene, this works well.

I'm writing this scene from Smith's POV, and the reader won't see the actions of Jones. However, I need to know them so I can keep my details straight for later in the book. This simple diagram is all I need. I can see the fight in its entirety and don't need to hunt back through my text.

If this scene is the climax, and we want Smith and Jones to have a fistfight, then we can:

- Add bad guys as needed to diminish ammunition.
- Adjust the number of times each character fires – multiple shots at bad guys diminishes ammunition.
- Wound characters.
- Break their guns.
- Maybe adjust weapons to limit round count. Revolvers generally have a six-round capacity. However, this may not be possible if you're restricted to an agency's firearm. It also has huge ripple effects for editing.

If you know the terrain where the action happens, you can play with any of the variables to achieve the desired result.

Note that I didn't specify the guns the random bad guys were using. None of them are important to this scene. However, they could be important to the story later. If so, have the characters identify the guns after the fight.

El Gato's gun is important. If Jones kills Tate with his own gun, a .45-caliber, then El Gato couldn't have killed Tate because the Glock 17 is a 9mm.

Admittedly, the scene outlined above ignores proper room-clearing tactics. That's another can of worms entirely. While those techniques are open source, I'll not be covering them in this guide.

Balancing the detail in a scene versus compromising military and police tactics is a tricky task.

Another thing, room clearing techniques and procedures vary wildly between agencies. What you write could get a quiet nod of recognition from one reader *AND* provoke a scathing review from another. Keep things vague enough to avoid this trap.

Distant Targets

What's Next in Firearms

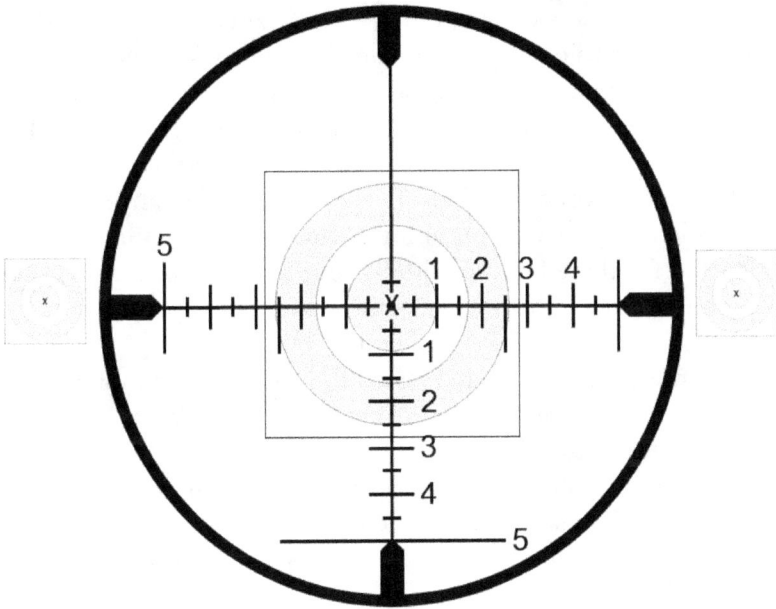

While it may seem that firearm technology has reached its zenith, there are still advances to be made. From the movie *Aliens*:

Ripley: Lieutenant, what do those pulse rifles fire?

Lieutenant Gorman: 10-millimeter explosive tip caseless. Standard light armor piercing rounds. Why?

Ah, the infantryman's dream, caseless ammunition – that or a brass magnet. Anyone who's shot on the range all day is well-versed in picking up brass at the end of the day. During this task, we would dream of caseless ammo. Is it really a dream?

The Rocket Ball was patented in 1848 by Walter Hunt. It consisted of a partially hollow projectile that contained the propellant. Since then, various companies have explored caseless ammunition. In 1961, Daisy, famous for its air rifles, had a .22-caliber caseless design that was shut down because it didn't have a license to produce firearms.

The Heckler & Koch G11 K is the closest any company has come to fielding a rifle that fires caseless ammo. How close? The German army had budgeted for three hundred thousand rifles in 1990. Unfortunately, political and economic factors came into play and the project was terminated.

In 1991, the company Voere created the VEC-91, a hunting rifle that used caseless ammo. This gun is no longer produced. Why?

It seems that the BATF and the US Congress decided that the advent of caseless ammunition represented a threat to citizen safety and the ability to link crimes back to a specific weapon. When the US cut off the market, Voere shut down production. Apart from some engineering issues, the only thing that prevents this evolutionary step forward is legislation.

The Voere is still available on some firearms auction websites. A great weapon for the discerning assassin, but rare. You'll need to do big research for this one.

Who doesn't love Judge Dredd? Fingerprint and DNA-activated guns are right around the corner, and fingerprint gun safes already exist. The only thing holding this technology back is processing speed. The lag between drawing the gun and being able to fire it is still too slow. A few seconds delay in a crisis can mean the difference between life and death.

I imagine governments and their agencies could mandate the use of this technology. However, good luck getting that past the gun lobbies. Mark my words though, the iGun is just around the corner.

Energy weapons are another advance that seems likely. Currently, directed energy weapons are too large for use as rifles or handguns, but so too were cannons at one time.

Plasma rifles fall into this category. If you do an internet search, you may be able to find garage tinkerers who've developed their own energy weapons. If a kid can build one in his basement using LED lights, you can be sure that an evil genius somewhere is working to perfect the design.

Gauss rifles are also becoming a real possibility. Also known as rail guns, weapons of this nature use electromagnetic coils to accelerate projectiles to high velocities. You can preorder yours today at Arcflash Labs for a cool $3,750. As with caseless ammo, I imagine BATF will have something to say about whether this technology ever becomes available.

All that said, the traditional firearm is likely to remain the go-to weapon for a long time. Legislation will probably be able to keep the genie in the bottle for a while. Cost and reliability will also play a role in the adoption rate of new weapons, just like they do with electric cars. Futuristic weapons are still a good way off.

However, when they do come, they will still need:

- A source of feed
- Grips
- A trigger mechanism
- Sights
- Safety/selector

These are all components your character will need to interact with to fire the new weapons.

Enough of Tomorrowland. Let's wrap this up so you can get back to writing.

The Parting Shot

Let me introduce you to the RR-36 Zero G Recoilless Rifle. It fires 20mm caseless ammunition and has a backblast tube for use in zero-g. It uses electrical ignition to fire the round.

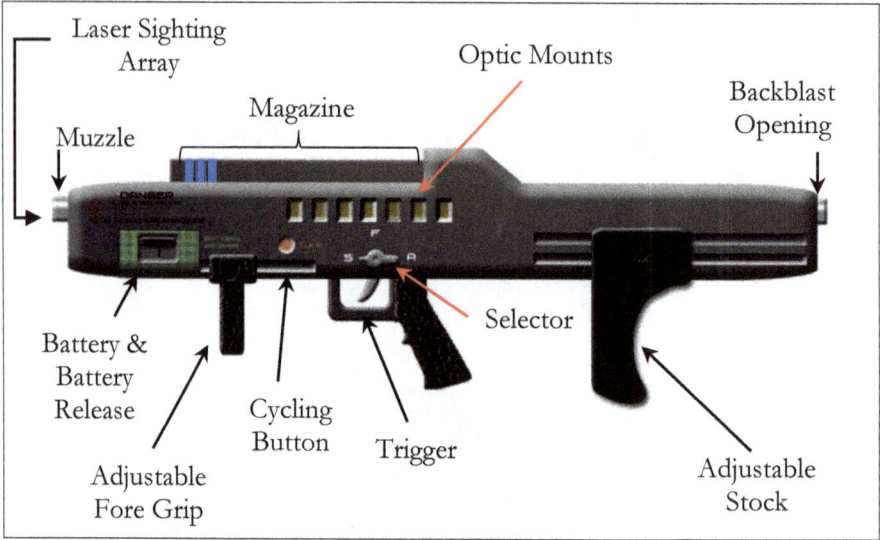

Laser Sighting Array
Optic Mounts
Backblast Opening
Magazine
Muzzle
Battery & Battery Release
Cycling Button
Trigger
Selector
Adjustable Fore Grip
Adjustable Stock

This is a fictitious rifle I designed for Scott Sigler to use in his novel, *The Crypt*. The book can be described as *Run Silent, Run Deep* meets *Event Horizon*.

Scott challenged me to create a weapon that could be used in space and wouldn't send the shooter spinning off into the void. Mr. Sigler could have waved his hands in the air and done that writer thing. You know, ignore physics. However, he is detail- and science-oriented, thus, the gun.

The RR-36 is a mashup of the RPG-7, H&K G11, and the FN P90. Just like the gunsmiths of old, I had to develop my own ammunition before I could develop the gun.

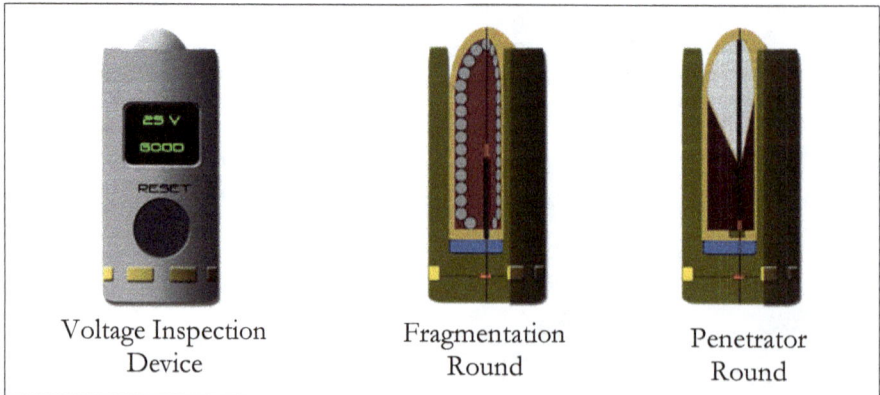

Voltage Inspection Device

Fragmentation Round

Penetrator Round

Then I thought of as many processes and contingencies as I could to include in the design. For example:

- **Load the gun** – Place the weapon on safe, insert the magazine, and press the cycle action button.
- **Unload the gun** – Remove magazine, press the cycle action button, and eject the chambered round.
- **Failure to fire** – Hit the cycling button to eject the bad round.
- **Second failure to fire** – Eject the battery and replace with the spare.

You're thinking, "Cool story, bro. What's the point?"

The point is that maybe 5 percent of this creation made it into the book. Like all real firearms, most of the detail is unimportant or too technical to include in an action scene.

Scott never used the failure to fire contingencies, but he might. He also knows that the gun is useless without a charged battery. What if the trooper's spare gets damaged? Well, then it's go time with the long knives.

You need to have the information at hand if you want to use it. From this guide you know that a character can't just pick up any rifle – equipped with iron sights – and effectively engage targets at a distance without zeroing it first. Likewise, you can't just slap on a new scope and expect accuracy without going to the range. You may never use that information, but then again, it could be a major plot point.

Balance your knowledge of guns, or any subject, with the needs of the story, plot, or character.

My advice is just that, advice. *You* are the god of the world you create. It's *your* story.

Thank you for reading my book. I hope it helps.

<<<<< Winchester >>>>>

Acknowledgments

This book would not exist if it weren't for the following people who either directly, or indirectly, influenced the course of my writing career.

Scott Sigler offered me a spot as a technical advisor on his book *Contagious*. It was my first gig as a TA, and it was a blast. I'm fortunate to call both Scott and his lovely wife, A, my friends. He made me write this book.

Kimberley Howe gave me my first speaking gig at the ThrillerFest conference in 2014. Some random guy dropped a form email through the ITW website, and she took a chance on me. I've spoken at the conference several times since then. Kimberley and the whole TF staff are amazing, generous people.

Chantelle Aimée Osman has given me more advice on writing and self-editing than you could possibly imagine.

Nathaniel Marunas contributed uncountable hours working on the design of the new cover, and he gave me pointers on improving the interior layout. Nathaniel taught me all the stuff you can't find online.

I'd like to extend a special thanks to all the writers who've asked me questions over the years. You are in these pages. E.G. Michaels, Matty Dalrymple, AJ Colucci, Michelle Daniel, Dwayne Goetzel, Jeanette Anderson, Julie Allen, Stacey Allen, Edwin Diaz, Elin Barnes, and all the others whose names escape me at this moment.

There's another group of people who deserve recognition, my beta readers: COL JP Harvey, Air Force (Ret) (a.k.a. the Bus Driver); Glenn Dyer (Author); and His Excellency, Richard Lesperance, former teammate, ODA 2072; and Kevin Mest, my plot hole guru. You guys made this work better, thank you.

Finally, I'd like to thank my wonderful wife, Mindy. Her patience and support are critical to my writing endeavors. She listens when I ramble on, asks important questions, and reminds me to breathe. She doesn't even complain too much when I stumble into bed after a late night of writing. Thanks, Sweetie, I love you.

Annex A: Full Character Sheet

Character:	
Manufacturer:	
Type and Action:	
Model:	
Caliber:	
Capacity:	
Size:	Full size = > 4.5" barrel Compact = 3.5 to 4.5" barrel Sub-compact = 3 to 3.5" barrel Pocket = Less than 3"
Safety:	
Suggested Carry Condition	
Condition when empty	
Clicks when empty*	
Historical Data	
Dates of production	
Military Use	
Character data	
Character Carry Condition	
Character Carry Location	
Staging Location (Home)	
Staging location (Car)	
Staging Location (Office)	
Gun Tags	
Notes and Identifying Features:	

Annex B: Firearm Baseball Card

Manufacturer:		
Type and Action:		
Model:		
Caliber:		
Capacity:		
Size:	Full size = > 4.5" barrel Compact = 3.5 to 4.5" barrel Sub-compact = 3 to 3.5" barrel Pocket = Less than 3"	
Safety:		
Suggested Carry Condition		
Condition when empty		
Clicks when empty*		
Gun Tags		

Annex C: Firearms Information Table

Name / Features				
Manufacturer				
Model				
Type				
Caliber				
Capacity				
Trigger Action				
Safety				
Magazine Release				
De-cock				
Slide release				
Clicks on empty				
Tags				

Annex D: Trigger Guard Firearm Baseball Cards

AK47

Ejection Port

Charging Handle

Safety/Selector
Colored blue
for visibility

Magazine Release

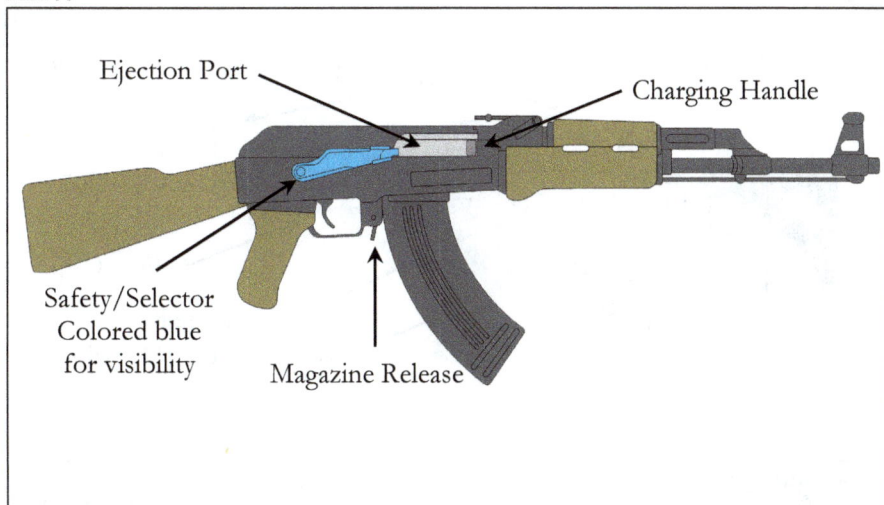

Manufacturer:	Various
Model:	47
Years of production/ Service	1948 – Present / 1949 – 1974 (Russia)
Type and Action:	Select fire, rifle (capable of full auto)
Caliber:	7.62 x 39mm
Capacity:	Usually 30-round magazine (Other sizes are available)
Safety:	Lever on the side of the receiver
Suggested Carry Condition	Loaded, on safe
Condition when empty	Bolt forward
Sound after final round is fired	CLICK Most AKs don't have a hold open feature. This gun will click once when empty
Notes	

AR-15 Variants

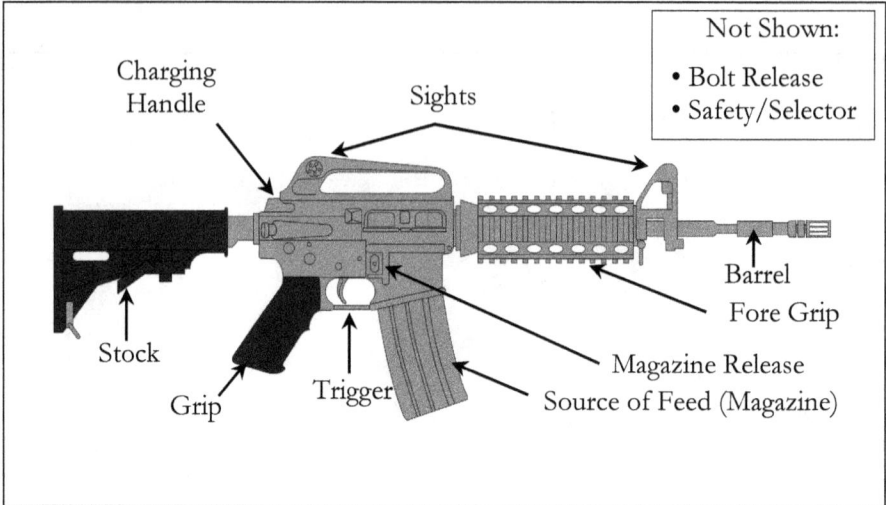

Charging Handle

Sights

Not Shown:
- Bolt Release
- Safety/Selector

Stock

Grip

Trigger

Barrel

Fore Grip

Magazine Release

Source of Feed (Magazine)

Manufacturer:	Various
Model:	AR-15, M16, M4
Years of production/ Service	1964 – Present M16: 1964 – Present (US) M4: 1994 – Present (US)
Type and Action:	Semi-automatic rifle (military versions are capable of firing in the fully automatic mode)
Caliber:	.223 Remington or 5.56 x 45mm NATO
Capacity:	Ejecting magazines are commonly 20-30-rounds but there are higher capacities available
Safety:	Selector switch on the receiver
Suggested Carry Condition	Round chambered, magazine in, on safe
Condition when empty	Bolt locks to the rear
Sound after final round is fired	None
Notes	AR stands for Armalite rifle The M16 A2 doesn't fire in the fully automatic mode. It fired 3-round bursts instead. Full-auto was re-introduced in the A3 model.

Beretta 92f

De-cocking Safety

Slide Release

Hammer

Trigger

Grip

Magazine

Slide locks on empty

Manufacturer:	Beretta
Model:	92f or 92fs // U.S. Military M9
Years of production/ Service	1985 – Present / M9 1985 – 2017 (US)
Type and Action:	DA/SA semi-automatic pistol
Sizes Available	Full
Caliber:	9mm Luger / Parabellum / Para
Capacity:	Usually 15 +1 (other magazine sizes available)
Safety:	De-cocking safety
Suggested Carry Condition	Loaded, hammer down, (Use of the safety is shooter preference or agency SOP)
Condition when empty	Slide locks to the rear
Sound after final round is fired	None
Notes	

Colt Cobra

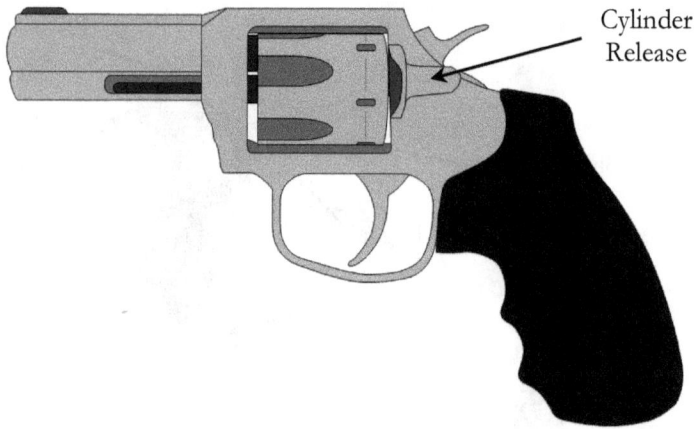

Cylinder Release

Manufacturer:	Colt
Model:	Cobra
Years of production/ Service	1950 – 1981, break in production, 2017 – Present
Type and Action:	Double action revolver (Hammer can be cocked for single action)
Sizes Available	Full, Compact, Subcompact
Caliber:	.38 Special
Capacity:	6-round cylinder
Safety:	NA
Suggested Carry Condition	Loaded, hammer down
Condition when empty	NA
Sound after final round is fired	Click. This gun is manually operated. It will click every trigger pull when empty
Notes	Like many firearms, this revolver is offered in multiple sizes.

Colt Single Action Army

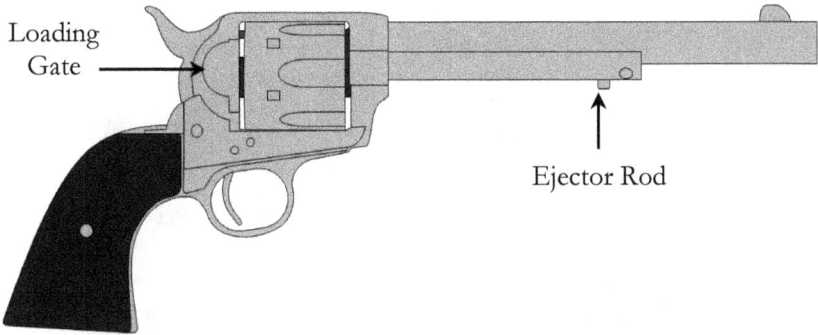

Loading Gate

Ejector Rod

Manufacturer:	Colt
Model:	NA (also known as the Peacemaker)
Years of production/ Service	1873 – Present (With periods of non production) / 1873 – 1945 (US)
Type and Action:	Single action revolver
Caliber:	Chambered in several calibers
Capacity:	6-round cylinder
Safety:	NA / Half-cock
Suggested Carry Condition	Full cylinder, half-cock
Condition when empty	NA
Sound after final round is fired	Click Revolvers are manually cycled
Notes	

Colt M1911

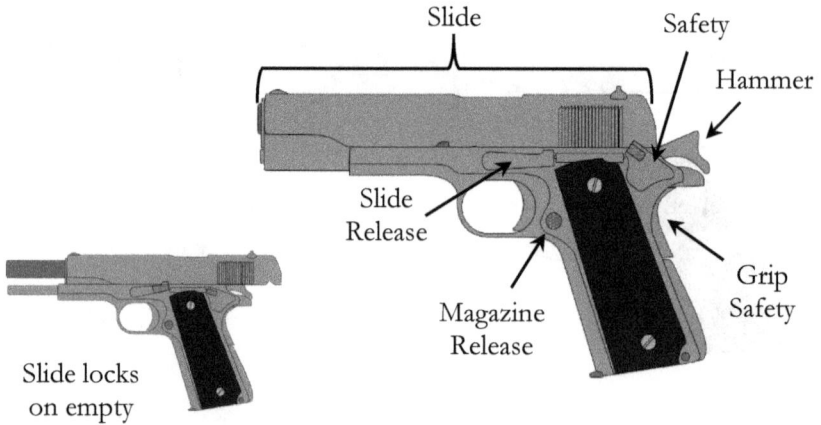

Slide — Safety — Hammer — Slide Release — Magazine Release — Grip Safety

Slide locks on empty

Manufacturer:	Colt and other manufacturers
Model:	1911
Years of production/ Service	1911 – Present / 1911 - Present
Type and Action:	Single action semi-automatic pistol
Sizes Available	Full
Caliber:	.45 ACP
Capacity:	7 +1 / Other magazine capacities available
Safety:	Yes
Suggested Carry Condition	Loaded, on safe
Condition when empty	Slide locks to the rear
Sound after final round is fired	None
Notes	The 1911 is manufactured by a lot of different companies. Each has their own variations in size and sight options, but the basic features seen here are common to all.

They are all known as 1911s. |

COP .357 Derringer

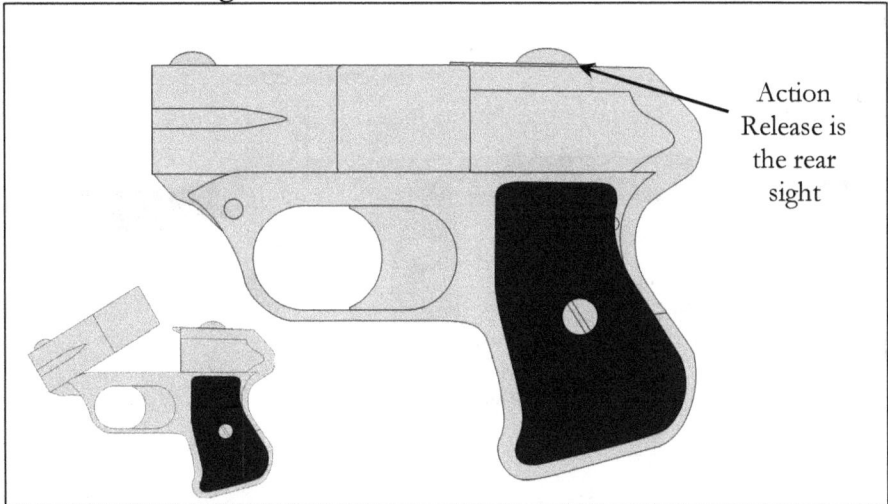

Action Release is the rear sight

Manufacturer:	COP Inc
Model:	NA
Years of production/ Service	1983 – 1990
Type and Action:	4-barreled derringer / Double action only
Caliber:	.357 Magnum / .38 Special
Capacity:	4 – individual barrels
Safety:	None
Suggested Carry Condition	Loaded
Condition when empty	NA
Sound after final round is fired	Clicks with each trigger pull
Notes	Terrible gun. I advise against using it.

FN FAL

Safety/Selector

Cocking Lever

Bolt Release

Magazine
Release

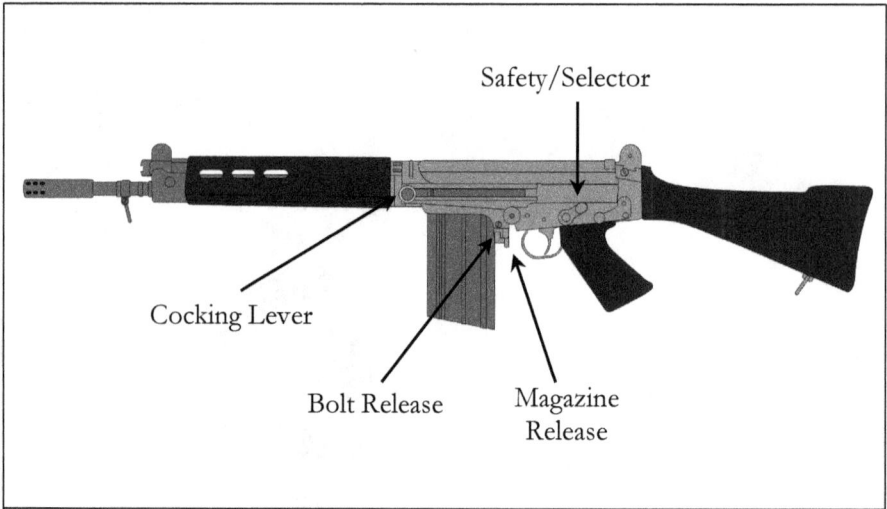

Manufacturer:	FN Herstal (Belgium)
Model:	FAL
Years of production/ Service	1953 – Present
Type and Action:	Select fire, rifle (capable of full auto)
Caliber:	7.62 x 51mm NATO
Capacity:	20 or 30-round ejecting box magazine
Safety:	Selector on the receiver
Suggested Carry Condition	Loaded, on safe
Condition when empty	Bolt locks to the rear
Sound after final round is fired	None
Notes	

FN Model 1910

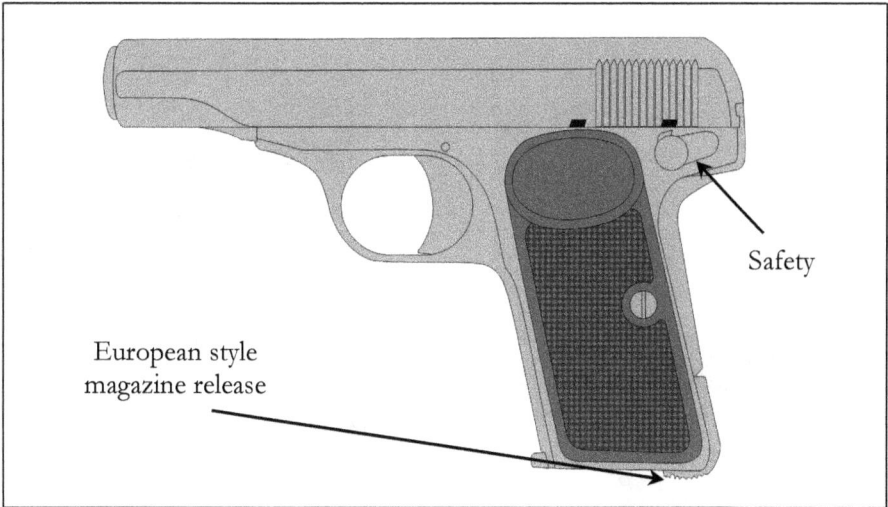

Safety

European style
magazine release

Manufacturer:	Fabrique Nationale (FN)
Model:	1910
Years of production/ Service	1910 - 1983
Type and Action:	Striker fired semi-auto pistol
Sizes Available	Compact
Caliber:	.380 ACP or .32 ACP
Capacity:	.380: 6-round ejecting magazine +1 .32: 7-round ejecting magazine +1
Safety:	None
Suggested Carry Condition	Loaded
Condition when empty	Slide forward
Sound after final round is fired	Will click one time
Notes	There is another model of this pistol, the Model 1922. Beware of firearms offered in multiple calibers.

The Character will need to work the slide during a reload. |

Glock 18

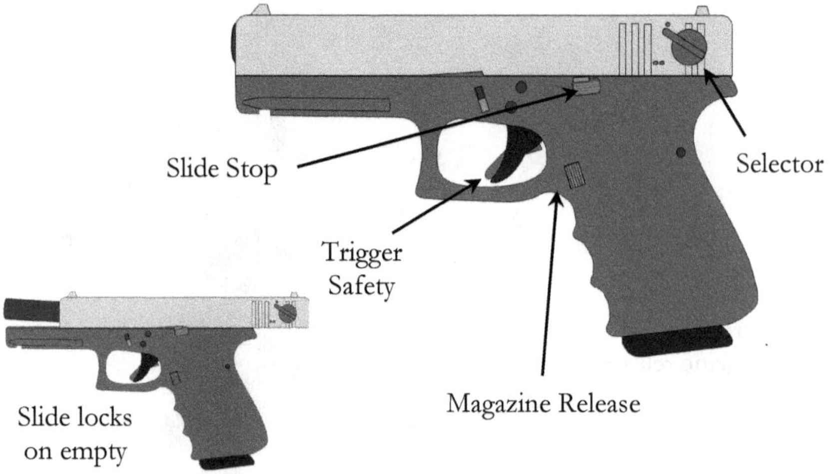

Slide Stop

Selector

Trigger
Safety

Slide locks
on empty

Magazine Release

Manufacturer:	Glock
Model:	G18
Years of production/ Service	1986 - Present
Type and Action:	Select fire machine pistol (Capable of full-auto fire)
Sizes Available	Full
Caliber:	9 x 19mm Parabellum
Capacity:	Various magazine sizes +1
Safety:	None
Suggested Carry Condition	Loaded
Condition when empty	Slide locks to the rear
Sound after final round is fired	None
Notes	This pistol is a modified Glock 17.

Glock 19

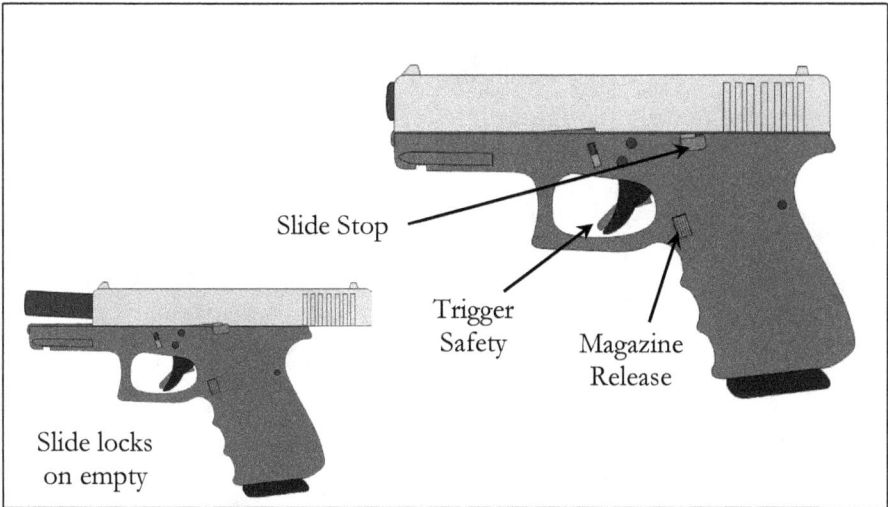

Slide Stop

Trigger
Safety

Magazine
Release

Slide locks
on empty

Manufacturer:	Glock
Model:	G19
Years of production/ Service	1982 – Present / 1982 – Present
Type and Action:	Striker fired semi-automatic Pistol
Sizes Available	Compact
Caliber:	9 x 19 Parabellum
Capacity:	15 +1 Ejecting magazine (Other magazine sizes available)
Safety:	None
Suggested Carry Condition	Loaded
Condition when empty	Slide locks to the rear
Sound after final round is fired	None
Notes	You've seen one Glock, you've seen them all. Remember that some Glock model numbers don't reflect the caliber of the pistol. Glocks are widely used in law enforcement.

Grendel P10

Special clip adapter needed
for loading by clip

Otherwise, rounds
are loaded
individually
through the top

Slide Release

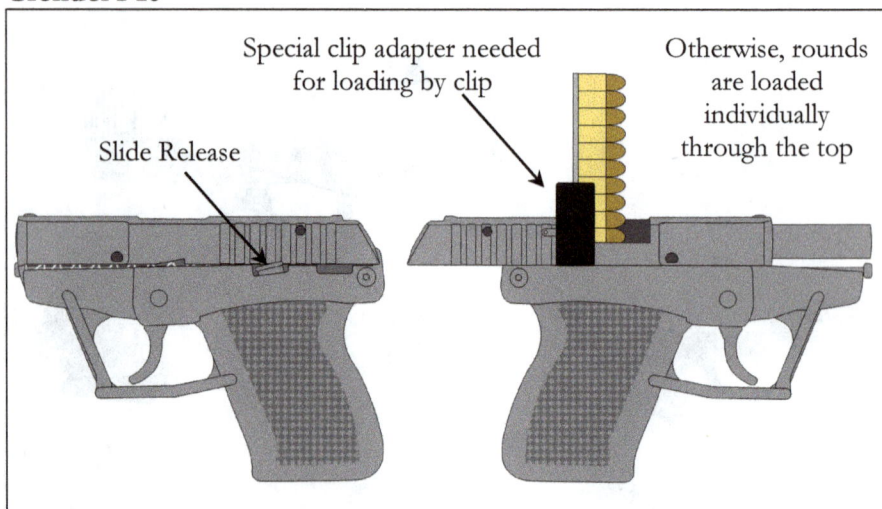

Manufacturer:	Grendel Inc
Model:	P10
Years of production/ Service	1988 – 1991
Type and Action:	Double action only semi-auto pistol
Sizes Available	Subcompact
Caliber:	.380 ACP
Capacity:	10-round internal magazine
Safety:	None
Suggested Carry Condition	Loaded
Condition when empty	Slide locks to the rear
Sound after final round is fired	None
Notes	Terrible gun. I advise against using it. The reloading procedure is messy and would need explanation.

H&K P30 Variant 1

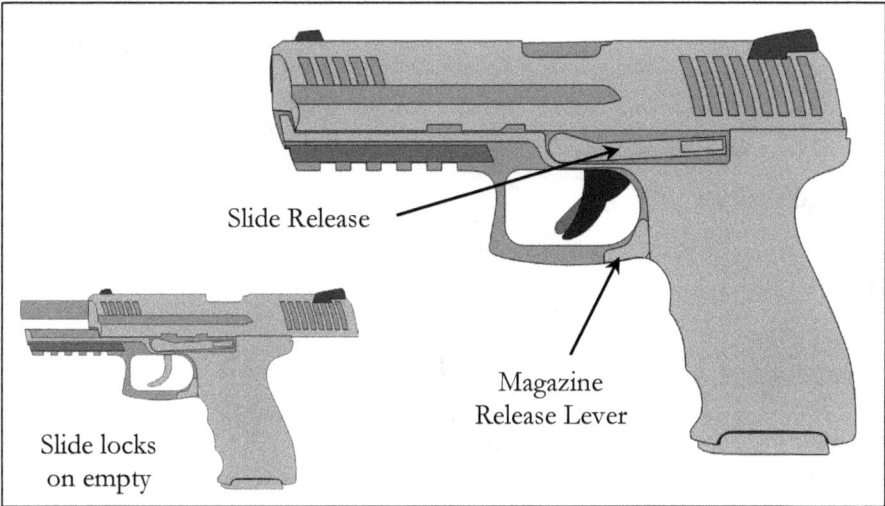

Slide Release

Magazine
Release Lever

Slide locks
on empty

Manufacturer:	Heckler & Koch
Model:	P30 Variant 1
Years of production/ Service	2006 - Present
Type and Action:	Double action only semi-auto pistol
Sizes Available	Full or Compact
Caliber:	9 x 19mm Parabellum & .40 S&W
Capacity:	Various magazines available +1
Safety:	None
Suggested Carry Condition	Loaded
Condition when empty	Slide locks to the rear
Sound after final round is fired	None
Notes	I wouldn't suggest this firearm due to possible variant confusion.

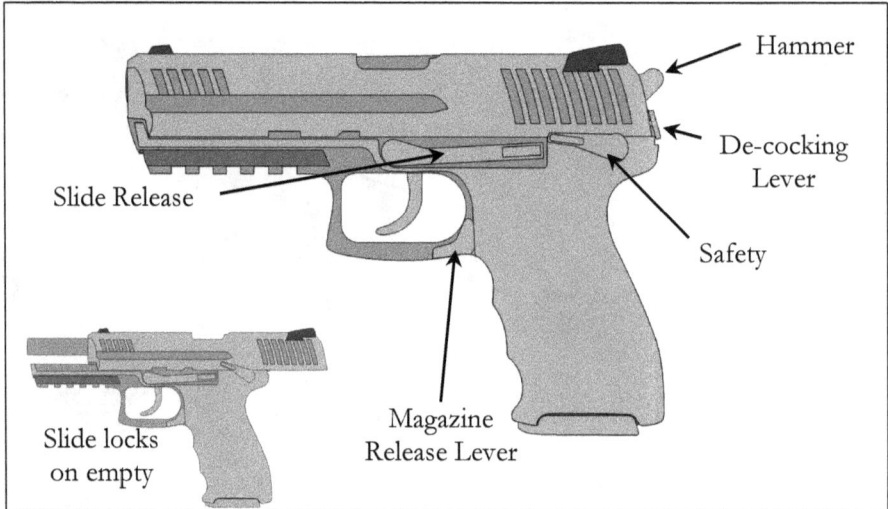

Hammer

De-cocking Lever

Slide Release

Safety

Slide locks on empty

Magazine Release Lever

Manufacturer:	Heckler & Koch
Model:	P30 Variant 3
Years of production/ Service	2006 - Present
Type and Action:	Double action semi-auto pistol (Hammer can be cocked
Sizes Available	Full or Compact
Caliber:	9 x 19mm Parabellum & .40 S&W
Capacity:	Various magazines available +1
Safety:	Yes / Separate de-cocking lever (Available with de-cocking lever only no safety)
Suggested Carry Condition	Loaded, hammer down. (Safety use by individual preference or agency SOP)
Condition when empty	Slide locks to the rear
Sound after final round is fired	None
Notes	I wouldn't suggest this gun due to variant confusion and the optional safety feature. Too many ways to trip up the writer.

Ingram Model 10

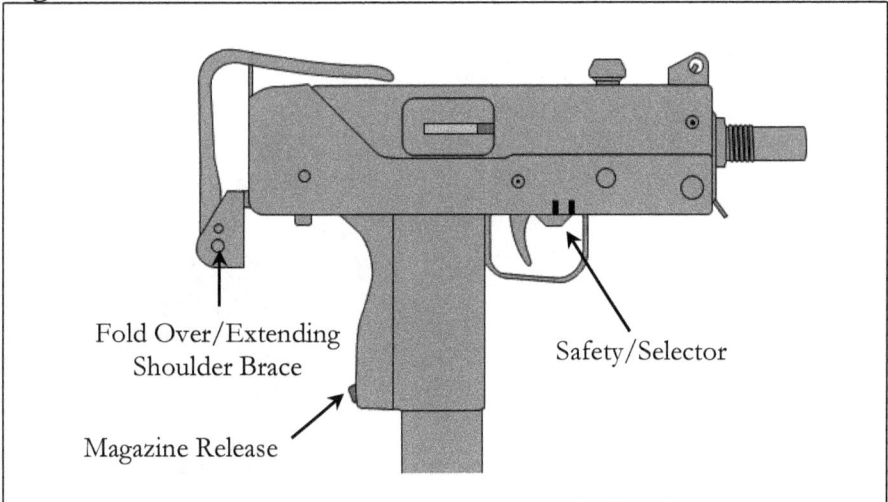

Fold Over/Extending
Shoulder Brace

Magazine Release

Safety/Selector

Manufacturer:	Military Arms Company
Model:	Model 10 (MAC-10)
Years of production/ Service	1970 - 1973
Type and Action:	Submachine gun / Machine pistol (Fires from the open bolt
Caliber:	.45 ACP
Capacity:	30-round ejecting magazine
Safety:	Selector slide near trigger or selector lever on the receiver
Suggested Carry Condition	Bolt to the rear, on safe
Condition when empty	Bolt goes forward
Sound after final round is fired	*Thunk* This firearm doesn't have a hold open feature
Notes	

Kimber K6s

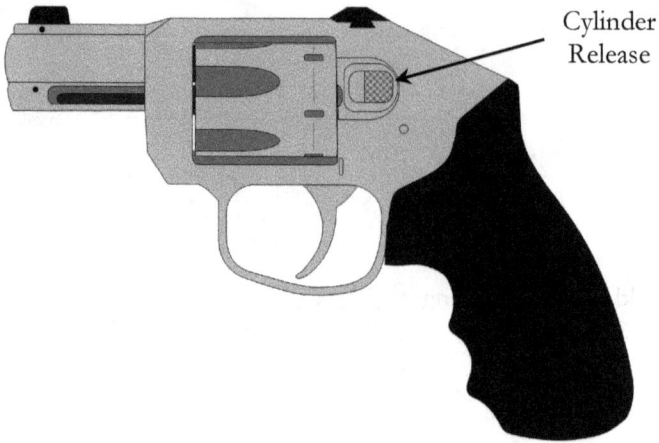

Cylinder Release

Manufacturer:	Kimber
Model:	K6s
Years of production/ Service	2016 – Present
Type and Action:	Double action only revolver (No hammer)
Sizes Available	Subcompact
Caliber:	.357 / .38 Special (can fire both types of ammo)
Capacity:	6-round cylinder
Safety:	NA
Suggested Carry Condition	Loaded cylinder
Condition when empty	NA
Sound after final round is fired	Click Revolvers are manually cycled
Notes	

Luger Pistol

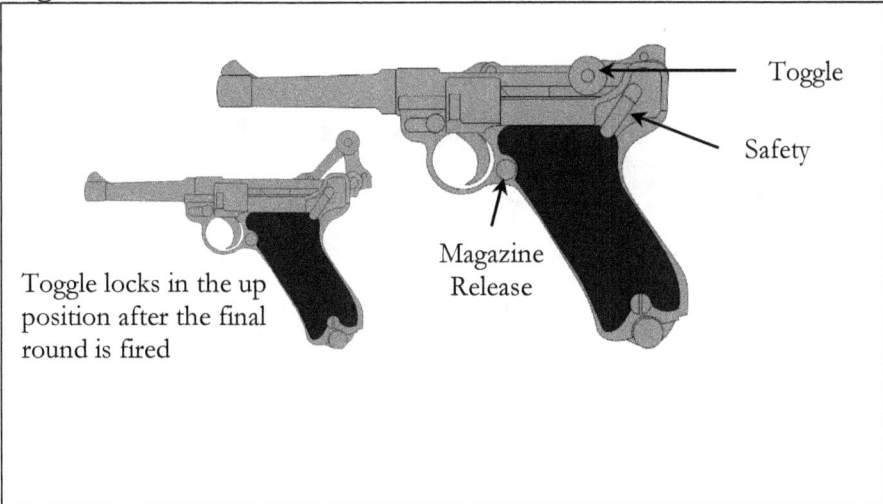

Toggle

Safety

Magazine
Release

Toggle locks in the up
position after the final
round is fired

Manufacturer:	Various manufacturers
Model:	P08
Years of production/ Service	1900 – 1953 / 1904 – 1953 (Germany)
Type and Action:	Semi-automatic striker fired pistol
Sizes Available	Full
Caliber:	7.65 x 21mm Parabellum & 9 x 19 Parabellum
Capacity:	8-round ejecting magazine
Safety:	Yes
Suggested Carry Condition	Loaded, (Safety use by individual preference or agency SOP)
Condition when empty	Toggle locks up
Sound after final round is fired	None
Notes	The Luger has multiple variants.

Marlin Model 336

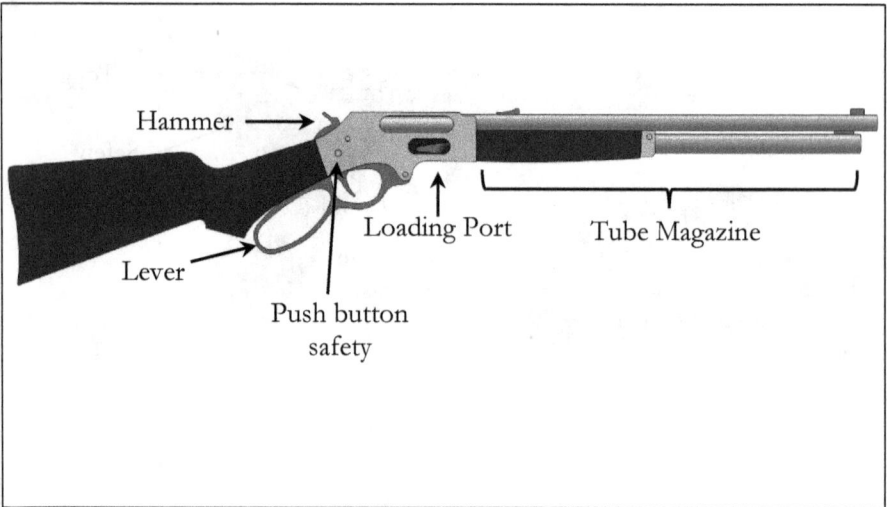

Hammer →

Lever →

Loading Port

Tube Magazine

Push button safety

Manufacturer:	Marlin / Remington
Model:	Model 336
Years of production/ Service	1948 – Present
Type and Action:	Lever action Rifle
Caliber:	Offered in a number of calibers
Capacity:	Varies by caliber
Safety:	Push button safety on the receiver
Suggested Carry Condition	Loaded and on safe
Condition when empty	NA
Sound after final round is fired	*Click* Can be cocked when empty
Notes	

Mauser C96

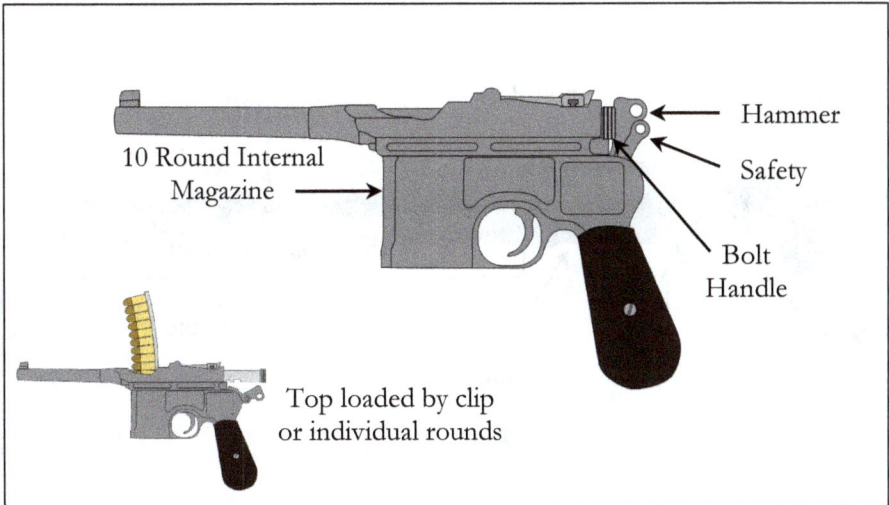

Hammer

10 Round Internal Magazine →

Safety

Bolt Handle

Top loaded by clip or individual rounds

Manufacturer:	Mauser and Various other companies
Model:	C96
Years of production/ Service	1896 – 1937 / 1896 – 1961 (Various countries)
Type and Action:	Single action semi-automatic pistol
Sizes Available	Full
Caliber:	7.63 x 25mm Mauser and other calibers
Capacity:	10-round internal magazine / capacity varies by caliber
Safety:	Yes
Suggested Carry Condition	Loaded, hammer back, on safe
Condition when empty	Bolt locks to the rear
Sound after final round is fired	None
Notes	There are multiple variants of this pistol.

Mossberg 500

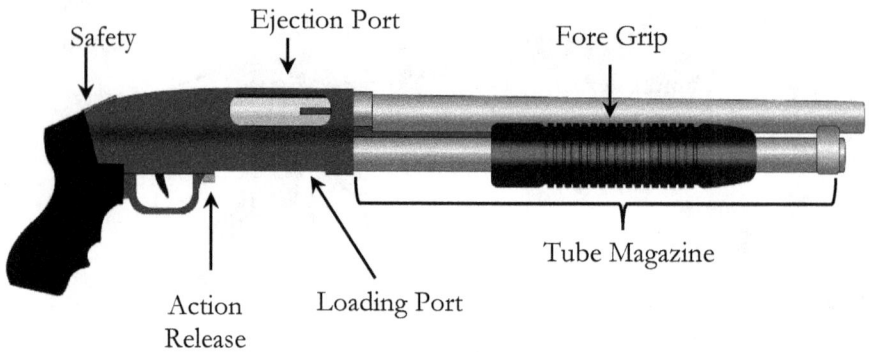

Safety — Ejection Port — Fore Grip — Tube Magazine — Action Release — Loading Port

Manufacturer:	Mossberg
Model:	500
Years of production/ Service	1961 – Present / 1987 – Present (US)
Type and Action:	Pump action shotgun
Caliber:	12-gauge (other gauges are available)
Capacity:	Varies by option (5 to 8 shells) / +1
Safety:	Slider on top of the receiver
Suggested Carry Condition	Loaded, on safe
Condition when empty	NA
Sound after final round is fired	*Click* Can be cocked when empty
Notes	Available in sporting versions

MP5

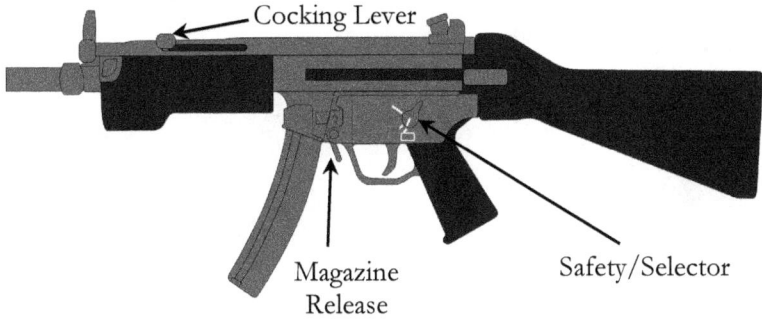

Cocking Lever

Magazine Release

Safety/Selector

Manufacturer:	Heckler & Koch
Model:	MP5
Years of production/ Service	1966 – Present
Type and Action:	Select fire submachine gun (capable of full auto)
Caliber:	9 x 19mm Parabellum
Capacity:	15, 30, 40, or 50-round magazines (+1)
Safety:	Selector lever
Suggested Carry Condition	Loaded on safe
Condition when empty	Bolt forward
Sound after final round is fired	*CLICK* – Because the bolt goes forward on an empty chamber
Notes	

M1 Garand

Loads from
the top

Clip Latch
(Other Side)

Operating
Rod

Safety

8 Round
Internal Magazine

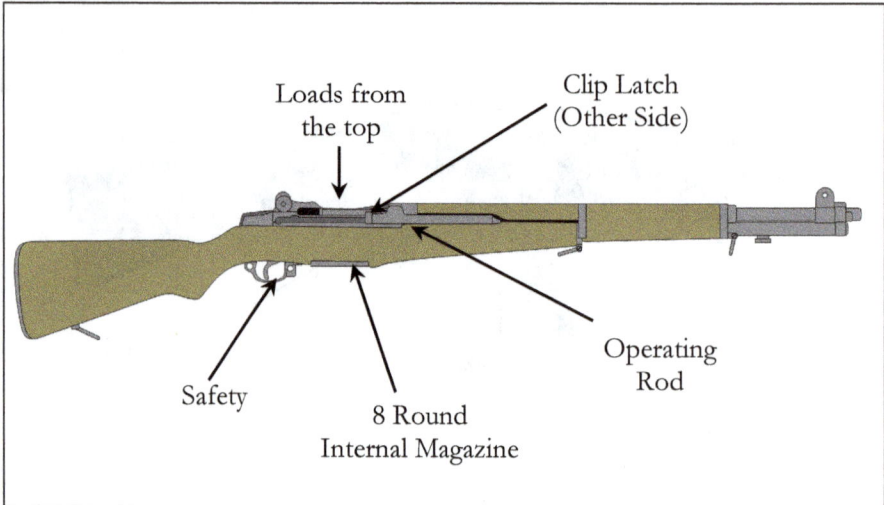

Manufacturer:	Various
Model:	M1 Garand
Years of production/ Service	1936 – Present / 1936 – 1957 (US)
Type and Action:	Semi-automatic rifle
Caliber:	.30-06 Springfield and 7.62 x 51mm NATO
Capacity:	8-round internal magazine
Safety:	Slider on the trigger guard
Suggested Carry Condition	Loaded, on safe
Condition when empty	Bolt locks to the rear
Sound after final round is fired	*PING* When the last round is fired the clip ejects with a loud ping
Notes	

M14

Selector

Operating Rod

Safety

Magazine Release

Manufacturer:	Various
Model:	M14
Years of production/ Service	1959 – Present / 1959 – Present (US) As primary rifle (US) 1957 - 1964
Type and Action:	Select fire, rifle (capable of full auto)
Caliber:	7.62 x 51 NATO
Capacity:	Usually 20-round magazine (other sizes available)
Safety:	Slider on trigger guard / Separate selector on receiver
Suggested Carry Condition	Loaded, on safe
Condition when empty	Bolt locks to the rear
Sound after final round is fired	None
Notes	

M60 Machine Gun

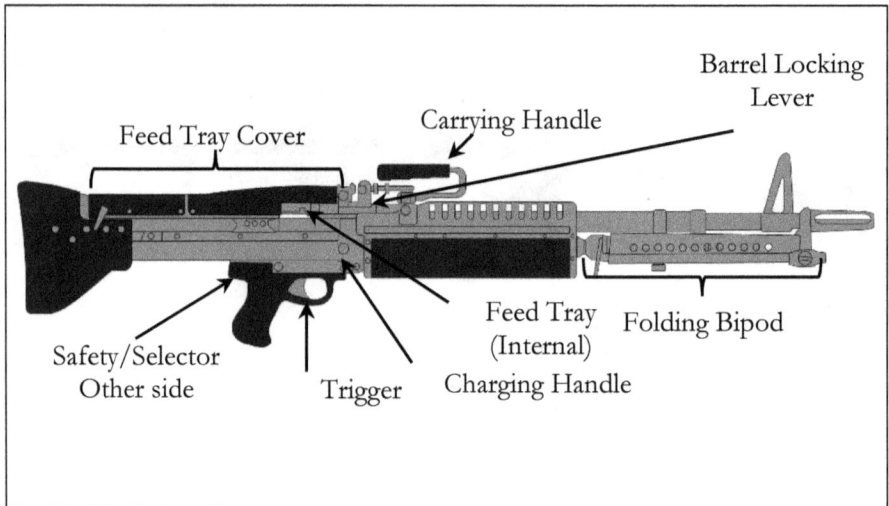

Barrel Locking
Lever

Carrying Handle

Feed Tray Cover

Feed Tray
(Internal)

Folding Bipod

Safety/Selector
Other side

Trigger

Charging Handle

Manufacturer:	U.S. Ordnance
Model:	M60 Machine gun
Years of production/ Service	1957 – Present / 1957 – Present (US)
Type and Action:	Fires from the open bolt
Caliber:	7.62 x 51mm NATO
Capacity:	Belt fed
Safety:	Switch near the trigger
Suggested Carry Condition	Bolt to the rear, belt inserted, on safe
Condition when empty	Bolt forward
Sound after final round is fired	*Thunk* – Because the bolt goes forward on an empty chamber
Notes	

Remington Model 870

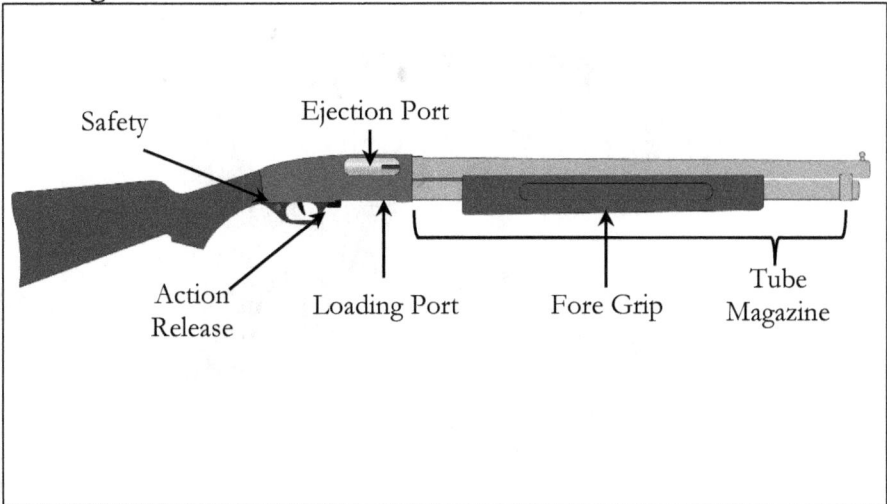

Manufacturer:	Remington
Model:	870
Years of production/ Service	1951 – Present / 1951 – Present (US)
Type and Action:	Pump action shotgun
Caliber:	12-gauge (other gauges are available)
Capacity:	Varies by option (3 to 7 shells) / +1
Safety:	Push button near the trigger
Suggested Carry Condition	Loaded, on safe
Condition when empty	NA
Sound after final round is fired	*Click* Can be cocked when empty
Notes	Available in tactical versions

Rohrbaugh R9

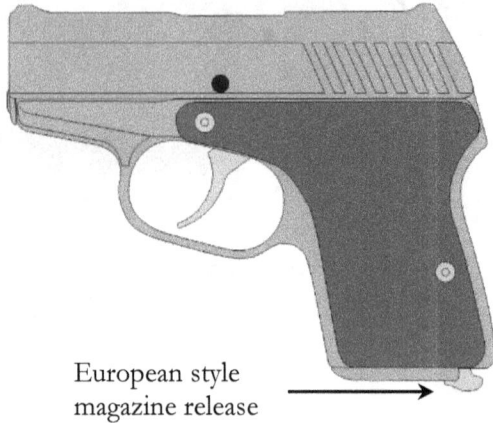

European style
magazine release ⟶

Manufacturer:	Rohrbaugh Firearms
Model:	R9
Years of production/ Service	2002 – 2014
Type and Action:	Double action only semi-auto pistol
Sizes Available	Pocket
Caliber:	9 x 19mm Parabellum
Capacity:	6-round ejecting magazine
Safety:	None
Suggested Carry Condition	Loaded
Condition when empty	Slide goes forward
Sound after final round is fired	Will click one time
Notes	The Character will need to work the slide during a reload.

Shadow Systems CR920

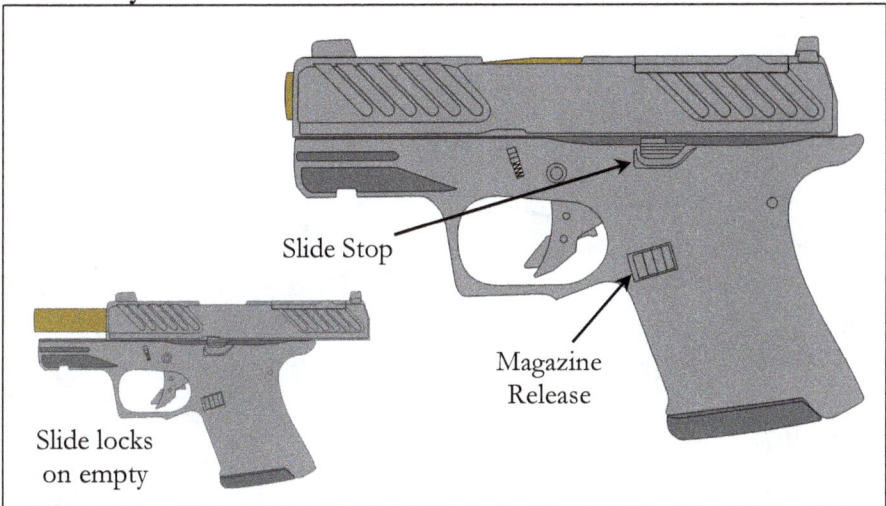

Slide Stop

Magazine
Release

Slide locks
on empty

Manufacturer:	Shadow Systems
Model:	CR920
Years of production/ Service	2016 – Present
Type and Action:	Striker fired semi-auto pistol
Sizes Available	Subcompact
Caliber:	9 x 19mm Parabellum
Capacity:	13 +1 ejecting magazine
Safety:	None
Suggested Carry Condition	Loaded
Condition when empty	Slide locks to the rear
Sound after final round is fired	None
Notes	The manufacturer's name isn't easily recognizable. Would suggest a different firearm.

SIG Sauer P226

Hammer

Takedown
Lever

De-cocking Lever

Slide
Release

Springs back into place
after de-cocking

Magazine
Release

Slide locks
on empty

Manufacturer:	SIG Sauer
Model:	P226
Years of production/ Service	1985 – Present / 1985 – Present
Type and Action:	Double action Semi-automatic pistol. (Hammer can be cocked for single action)
Sizes Available	Compact
Caliber:	9 x 19 Parabellum (Other calibers available)
Capacity:	15 +1 Ejecting magazine (Other magazine sizes available)
Safety:	NO, De-cocking lever returns the gun to double action mode
Suggested Carry Condition	Loaded, hammer down
Condition when empty	Slide locks to the rear
Sound after final round is fired	None
Notes	This gun is widely used in law enforcement.

Skorpion VZ 61

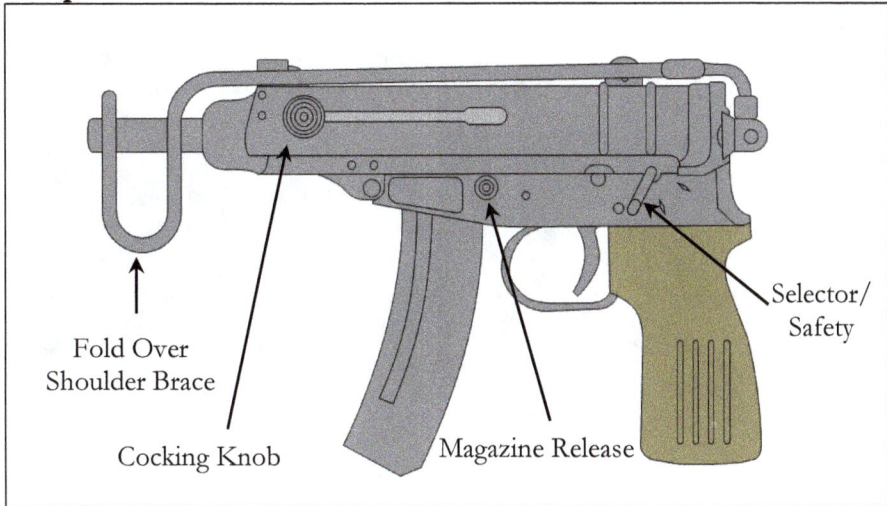

Fold Over
Shoulder Brace

Cocking Knob

Magazine Release

Selector/
Safety

Manufacturer:	Brod, Zastava Arms
Model:	VZ 61
Years of production/ Service	1961 – 1979 / 1961 - Present
Type and Action:	Select fire machine pistol (Capable of full-auto fire)
Caliber:	.32 ACP / 9mm variants exist
Capacity:	10, 20-round magazine +1
Safety:	Selector on the receiver
Suggested Carry Condition	Loaded, on safe
Condition when empty	Bolt locks to the rear
Sound after final round is fired	None
Notes	No bolt release. Work the cocking knob during reloads.

Steyr AUG

Cocking Slide

Firing Chamber
(Internal)

Ejection Port
(Other Side)

Folding Grip

Trigger

Safety/Selector

Bolt
Release

Magazine
Release

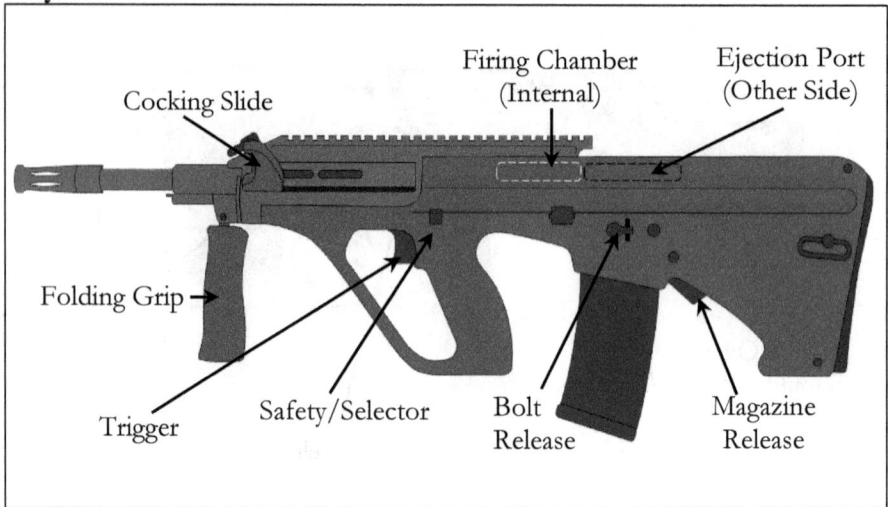

Manufacturer:	Various
Model:	Sometimes known as StG 77
Years of production/ Service	1977 – Present
Type and Action:	Bullpup Select fire, rifle (capable of full auto)
Caliber:	5.56 x 45mm NATO or 9 x 19mm Para
Capacity:	Varies by type +1
Safety:	Selector on the receiver
Suggested Carry Condition	Loaded, on safe
Condition when empty	Bolt locks to the rear
Sound after final round is fired	None
Note	This firearm is modular and can have multiple configurations.

Thompson Submachine Gun

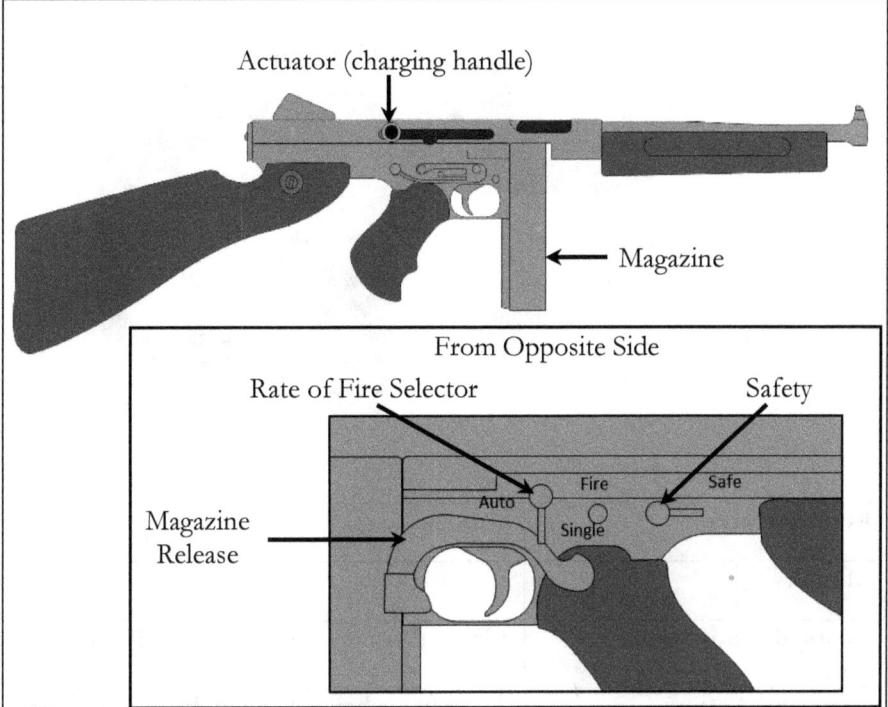

Actuator (charging handle)

Magazine

From Opposite Side

Rate of Fire Selector

Safety

Fire

Safe

Auto

Single

Magazine
Release

Manufacturer:	Various
Model:	M1921
Years of production/ Service	1921 – 1945 / 1921 – 1971 (US)
Type and Action:	Select fire submachine gun. Fires from open bolt
Caliber:	.45 ACP
Capacity:	20 and 30-round magazines. 100-round drum
Safety:	Lever on the receiver / Separate selector in full auto versions
Suggested Carry Condition	Bolt back, on safe
Condition when empty	Bolt forward
Sound after final round is fired	*Thunk* – Because the bolt goes forward on an empty chamber
Notes	New replicas fire from the closed bolt

Uzi

Cocking Knob

Selector/Safety

Magazine
Catch

Grip Safety

Magazine

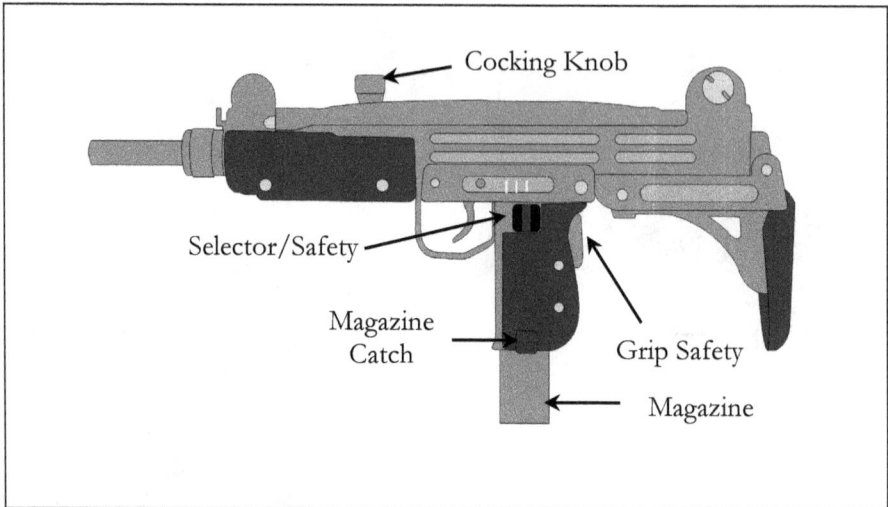

Manufacturer:	Various
Model:	
Years of production/ Service	1954 – Present / 1954 – Present (Israel)
Type and Action:	Select fire submachine gun. Fires from open bolt (full auto models)
Caliber:	Various. Usually 9 x 19 Parabellum
Capacity:	Various sizes available. Capacities vary by caliber.
Safety:	Selector slider near trigger
Suggested Carry Condition	Bolt back, on safe
Condition when empty	Bolt forward (Full auto models)
Sound after final round is fired	*Thunk* – Because the bolt goes forward on an empty chamber
Notes	This gun has multiple variants.

The civilian model fires from the closed bolt. These will make no sound when empty |

Vepr-12

Charging Handle

AK style safety/
Selector

Bolt Release

Magazine
Release

Ejecting Magazine

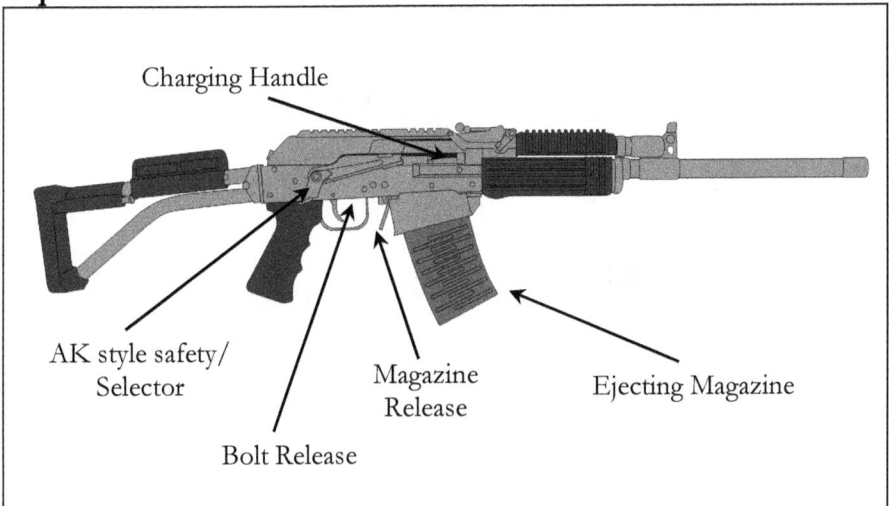

Manufacturer:	Molot Weapons (Russia)
Model:	Vepr-12
Years of production/ Service	2003 – Present
Type and Action:	Semi-auto shotgun
Caliber:	12-gauge
Capacity:	2 to 12-round magazines are available (+1)
Safety:	AK style lever / Selector
Suggested Carry Condition	Loaded, on safe
Condition when empty	Bolt locks to the rear
Sound after final round is fired	None
Notes	

Walther PP

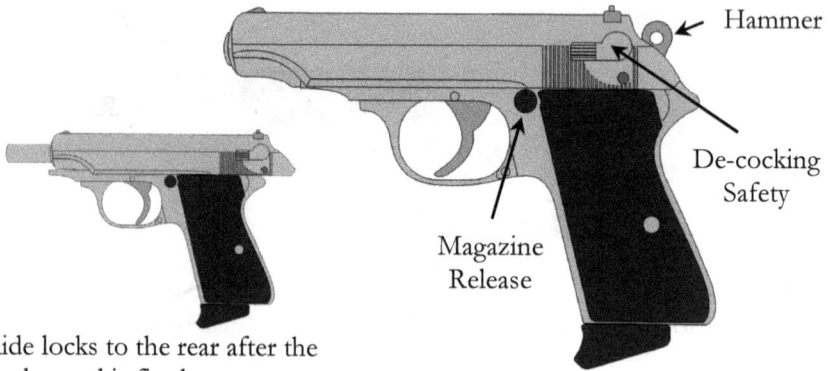

Hammer

De-cocking
Safety

Magazine
Release

Slide locks to the rear after the
final round is fired

Manufacturer:	Walther
Model:	PP
Years of production/ Service	1929 – Present / 1929 – Present
Type and Action:	Double action Semi-automatic pistol. (Hammer can be cocked for single action)
Sizes Available	Compact
Caliber:	.380 ACP (Other calibers available)
Capacity:	7-round magazine for .380-caliber (Varies for other calibers)
Safety:	De-cocking safety
Suggested Carry Condition	Loaded, hammer down. (Safety use by individual preference or agency SOP)
Condition when empty	Slide locks to the rear
Sound after final round is fired	None
Notes	There are many variants of this pistol. Most notable, the PPK. (James Bond's gun)

Webley-Fosbery

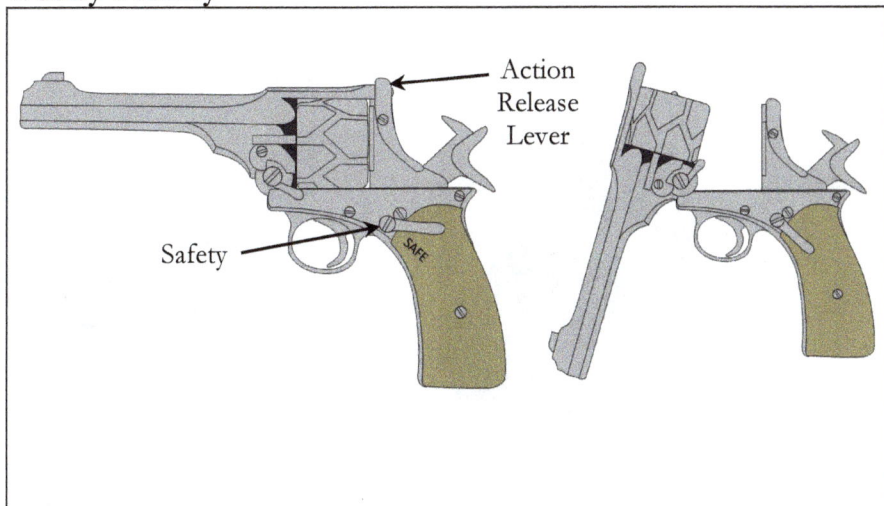

Manufacturer:	Webley (British)
Model:	NA
Years of production/ Service	1901 – 1924
Type and Action:	Automatic revolver. Single action trigger
Caliber:	.455 Webley MK II & .38 ACP
Capacity:	.455 – 6 round cylinder .38 – 8 round cylinder
Safety:	Yes
Suggested Carry Condition	Loaded, on safe
Condition when empty	NA
Sound after final round is fired	Click. This gun will click once when empty. The operator would need to cock the hammer for a second click.
Notes	

Webley Mark 1

Action Release
Lever

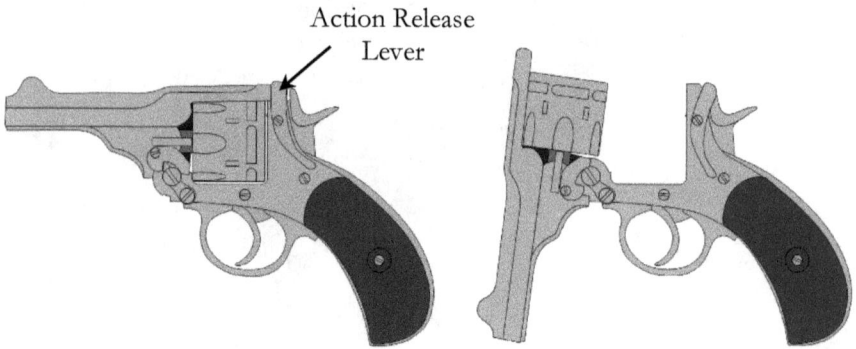

Manufacturer:	Webley (British)
Model:	Mark 1 – Mark 4
Years of production/ Service	Mark 1: 1887 – 1925 Mark 4: 1932 – 1978 / 1887 – 1978 (UK)
Type and Action:	Double/Single action revolver
Caliber:	Chambered in several calibers
Capacity:	6-round cylinder
Safety:	NA
Suggested Carry Condition	NA
Condition when empty	NA
Sound after final round is fired	Click Revolvers are manually cycled
Gun Tags	

Blank

Manufacturer:	
Model:	
Years of production/ Service	
Type and Action:	
Sizes Available	
Caliber:	
Capacity:	
Safety:	
Suggested Carry Condition	
Condition when empty	
Sound after final round is fired	
Gun Tags	

Glossary

accidental discharge: Also, negligent discharge. The act of firing the gun unintentionally. This occurs when the gun is improperly handled. This includes failing to properly clear the weapon.

ACP: Abbreviation for automatic Colt pistol. This is a caliber designation that refers to the headspace of the round.

action: A term used to describe how the firearm is operated or the operations of a trigger.

action, automatic: Any firearm that performs the tasks of extracting, ejecting, and loading of ammunition without any work required by the shooter. This class of weapons is further broken down by the ability, or inability, to fire multiple rounds with one pull of the trigger.

The term *automatic* is commonly used to describe fully automatic weapons. As such, always use semi-automatic to describe firearms that can't fire multiple rounds with one trigger pull.

action, automatic, full: Full-auto weapons that are capable of continuous fire as long as the trigger is fully depressed.

action, automatic, semi: Semi-auto firearms that require the trigger to be pulled for each round fired.

action, bolt: A type of rifle where the bolt is manually operated by an attached handle. Working the bolt to the rear extracts and ejects a spent casing. As the bolt is worked forward, a fresh cartridge is placed into the chamber, and the trigger is reset.

action, break: A type of firearm where the chamber(s) is/are accessed by releasing the barrel or chamber assembly to hinge downward, allowing the shooter to eject spent casings and load fresh cartridges.

action, double (pistol): A term used to describe actions performed by the pistol's trigger. Double-action triggers draw back the hammer or striker (the first action) and release it (the second action). Firing the pistol.

action, double/single (pistol): A type of trigger that acts as a double-action trigger for the first shot in semi-automatic pistols. When the first shot is fired, the weapon cycles and re-cocks the hammer or striker. The trigger is now in single-action mode for all subsequent shots until the hammer is de-cocked.

333

action, lever: A type of rifle, pistol, or shotgun where the bolt is manipulated by a lever, usually found on the underside of the gun. As the lever is pushed away, spent casings are extracted and ejected. When the lever is returned, a fresh cartridge is loaded, and the trigger is reset.

action, pump: A type of rifle or shotgun where the bolt is manipulated by a sliding fore grip. As the fore grip is drawn back, the spent cartridge is extracted and ejected. When the fore grip is sent forward, a fresh cartridge is loaded into the chamber.

action, single (pistol): Any semi-automatic pistol or revolver where the trigger performs the sole task of releasing the hammer or striker. In the case of revolvers, the hammer must be cocked each time by the operator.

ammunition: A combination of propellant and projectile combined in a single package or cartridge.

ammunition, subsonic: Ammunition designed to propel a bullet at a velocity slower than the speed of sound.

arquebus: An early musket with a hook or lug on the fore-end of the gun. The hook was used in conjunction with wall fortifications or a fork rest.

AR rifle: An ArmaLite rifle, originally designed by Eugene Stoner and sold to Colt Firearms company. M16 variants, M4 variants, and rifles using the Stoner design are all considered ARs.

assault rifle: A poorly defined political term generally meaning a magazine fed, automatically operated rifle capable of semi-automatic, burst, and/or automatic fire. Rate of fire is determined by a selector lever or switch. This definition may change between jurisdictions.

backup iron sights (BUIS): A set consisting of front and rear sights that can be mounted on a rifle. Sights of this nature are installed as a backup to optical systems that may fail when their batteries run out.

ball ammunition: See bullet, full metal jacket.

ballistics: The study of projectile behavior from launch to impact with a target. When applied to firearms, there are four distinct fields of study. They are listed in chronological order below.

ballistics, external: The study of projectile behavior while in flight.

ballistics, internal: The study of projectile behavior while it's in the barrel of a firearm.

ballistics, terminal: The study of projectile behavior as it impacts a target.

ballistics, transitional: The study of projectile behavior from the point when it leaves the barrel to the point when pressures from expanding gas from the propellant no longer influence its flight.

barrel: The part of a firearm that contains expanding gas from the propellant as it drives the projectile forward at high velocity.

battery, in: The condition of the firearm where the bolt or slide is fully forward and locked. Firearms *in battery* can be fired. Firearms *not in battery* can't be fired.

birdshot: A class of shotgun ammunition consisting of tiny balls or pellets ranging in size from .08 to .22 inches.

blank: A cartridge consisting only of primer and propellant.

bolt: The part of a firearm designed to seal the breech during firing. The bolt generally houses the firing pin and extractor. Some firearms incorporate the ejector as part of the bolt.

bore: The interior of the gun barrel.

brass: A common term for an expended shell casing.

breech: The part of the barrel open for the loading of ammunition.

buckshot: A class of shotgun ammunition consisting of metal balls ranging in size from .24 to .39 inches.

bullet, armor-piercing: A projectile with a hardened metal core encased in a softer jacket.

bullet, dum-dum: A bullet designed for maximum expansion on impact with a target. This term is archaic and seldom used in the modern era.

bullet, frangible: A bullet designed to disintegrate when it hits a hardened target.

bullet, full metal jacket: A soft core, usually lead, projectile fully encased in a harder metal jacket. This jacket prevents deformation of the bullet during the loading process and allows for higher muzzle velocity. Bullets of this type are known to overpenetrate or pass through a target. Also known as ball ammunition. Commonly abbreviated as FMJ.

bullet, hollow-point: A projectile with an exposed cavity at the nose. Projectiles of this type are designed for maximum expansion to cause increased damage and decrease the possibility of over penetration.

bullet, semi-jacketed: A projectile in which the metal jacket does not fully encapsulate the core.

bullet, wadcutter: A projectile specifically designed to shoot paper targets.

BUIS: Backup iron sights. A sighting system that is used when an optic fails.

bullpup: A rifle design in which the trigger is located in front of the ejection port. Rifles with this design can have a longer barrel and a shorter overall length.

burst: A string of shots fired from a fully automatic weapon. Example: a six- to nine-round burst. Firing in this manner gives the shooter greater control over recoil and helps with barrel temperature management.

burst mode: A select fire option in some automatic weapons that fires a set number of rounds with a single pull of the trigger.

butt: The back of a rifle's stock that rests against the shoulder. Also, the bottom of a pistol's grip.

caplock: Also, percussion lock. A lock mechanism for firing a musket.

caliber: The dimensions of a bullet and cartridge. The numerical portion is expressed in imperial units as hundredths of an inch, and in metric as millimeters. Secondary information describes the dimensions of the firearm's chamber or bore, which the round can safely be fired from/through. Examples: .45 ACP, .357 Magnum, .30-06 Springfield.

carbine: A shorter-barreled version of a standard rifle.

cartridge: An entire assembly of projectile, primer, and propellant. Musket cartridges were made of paper and only contained projectile and propellant.

> **cartridge, centerfire:** A cartridge of ammunition with its primer in the center of the base of the casing.

> **cartridge, magnum:** A cartridge designed with more propellant and a heavier projectile than a cartridge of similar caliber.

> **cartridge, paper:** A form of ammunition used with muzzle-loading muskets. The paper cartridge held enough powder and shot for a single firing round.

> **cartridge, paper, integrated:** A form of ammunition used with early breech-loading firearms. The integrated paper cartridge held the primer, propellant, and projectile(s).

> **cartridge, pinfire:** An early form of integrated metallic cartridge that contained the firing pin.

> **cartridge, rimfire:** A cartridge with its primer contained in the rim of the casing.

case/casing: The part of the cartridge that houses the propellant, primer, and projectile.

chamber: The part of a barrel designed to fit a specific cartridge. Revolvers and some shotguns have multiple chambers.

charging handle: A device used to manipulate the bolt. In automatic weapons it's used to manually cycle the action.

choke: A device mounted inside the barrel of a shotgun, designed to alter the distribution of shot as it leaves the muzzle.

clear: (v.) The unloading of a firearm. (adj.) unloaded.

clip: A metal strip designed to hold cartridges by the rim until they are ready to be loaded into a magazine.

cock: The act of drawing back the hammer to a firing position. In muskets, the portion of the lock that holds a slow match or flint.

cylinder: A device that contains multiple chambers. This can be revolved manually or mechanically depending on the model of revolver.

derringer: A small pistol with one or more barrels; designed to be a concealed weapon.

discharge: The act of firing a gun.

D.O.P.E.: 1. Data on Personal Equipment – The sum of adjustments from a mechanical zero specific to a single shooter and their weapon. 2. Data on Previous Engagement – Data a sniper records in their logbook.

double barrel: A rifle or shotgun with two barrels mounted side by side. Double barrel weapons may be equipped with two triggers or a barrel selector switch/lever.

double feed: A stoppage or jam caused by two rounds attempting to enter the chamber at the same time.

double tap: Two shots fired quickly.

dry: An unloaded firearm. Can be used to describe a firearm that has been fired until empty, specifically in thought or dialogue.

dry fire: The act of practicing with an unloaded firearm.

ejection: The process of clearing a cartridge from inside the firearm. This task is accomplished by the ejector.

ejection port: A window in the frame of the gun that allows spent casings to be expelled.

ejector: A component of the bolt or slide face that ejects the cartridge or cartridge case from the firearm.

experienced recoil: Also known as kick, the amount of force the shooter feels when the gun is fired. Recoil springs absorb some of this force reducing the effects on the shooter. The recoil spring is also an important operational part of automatic firearms.

extraction: The process where a cartridge or cartridge case is removed from the firing chamber. This task is accomplished by the extractor.

extractor: A component of the bolt or slide that grips the cartridge by the rim and removes it from the chamber when the action is worked.

eye relief: The distance between a magnified scope and the eye where the shooter has the maximum field of view the optic can provide.

fanning: A quick draw technique for single action revolvers where the shooter draws the revolver with the trigger pulled and fans the hammer with the off hand. Can be used to fire multiple rounds quickly.

field strip: The partial disassembly of a firearm for cleaning and maintenance. This level of disassembly does not require tools.

firearm: Any weapon that propels a projectile down a barrel by burning a propellant.

firing pin: A component of the bolt or slide that impacts the primer on a chambered cartridge. In some firearms, the firing pin may be held in place, under spring tension, by a sear that is released when the trigger is pulled. Other firearms use a hammer to strike the firing pin, forcing it forward to strike the primer.

flash suppressor: A device on the end of a barrel designed to reduce the visible signature of a weapon as it fires.

flier: An errant part of a shot group, not part of the main grouping. In a string of five shots, if four strikes are tightly grouped and the fifth is a noticeable distance away, it's called a flier.

flintlock: A device used to fire a musket. The cock was fitted with a piece of flint. When the trigger was pulled, the flint would swing forward and impact the frizzen. This created sparks that would ignite the primer charge in the pan.

frame: The part of the pistol or revolver that houses all internal components of the firearm. Similar to the receiver in rifles.

frizzen: The steel part of a flintlock firing system.

full cock: The final position of the hammer before firing.

full metal jacket (FMJ): See bullet, FMJ

Garand thumb (a.k.a. M1 thumb): an injury inflicted on the operator when the loading thumb is smashed by the bolt during the loading procedure. This injury is specific to the M1 Garand.

gauge: The measurement of the diameter of a shotgun's bore.

ghost gun: A firearm with no serial number; assembled from a kit, machined parts, or 3-D printed parts.

grain (gr): A unit of measure used to weigh bullets. 1 gr equals 1/7,000th lb.

grip safety: A mechanical safety on the grip of a firearm that must be depressed in order to fire.

grooves: The recessed portion of a rifled barrel.

half cock: A safety position of the hammer or cock that allows for safe loading. This position doesn't permit the trigger to be pulled.

hammer: The part of the gun designed to fire the weapon by either striking a primer directly, or by striking a firing pin, which, in turn, strikes the primer.

hammerless: Any pistol or revolver without an external hammer that could be manipulated by the operator.

holographic sight (holo sight): An unmagnified optical targeting system that projects an aiming point onto the glass. Sights of this type allow for faster engagement of targets.

iron sights: A simple sighting system consisting of a front and rear sight. They can be made of any material, but the historic name is what's used.

jacket: The hard metal coating that surrounds the core of a bullet.

jam: A common term used to describe an unintentional break in the firing cycle. A jam can be either a stoppage or malfunction.

kick: A common term used to describe recoil experienced by the shooter when the gun is fired. See experienced recoil.

lands: The raised portion of a rifled barrel.

load: (v.) The act of preparing a gun to fire. (n.) the weight of the bullet or the amount of propellant used.

M1 thumb: See Garand thumb.

machine gun: A firearm specifically designed to be fired in a fully automatic mode.

malfunction: An unintended break in the firing cycle caused by mechanical failure. Malfunctions may be caused by poor maintenance, lack of lubrication, damage to the weapon, or failure of an operational part.

magazine: An internal or external device used to house ammunition and present the next cartridge for loading in manually or automatically operated weapons. Magazines can be tube or box.

magnum: See cartridge, magnum.

manual safety: A mechanical safety that can be visually confirmed as engaged.

matchlock: A device used to fire a musket. It used a slow match to ignite the primer charge contained in an external pan.

MOA: Minute of angle. An angular measurement used to adjust sighting systems on firearms. 60 MOA = 1 degree.

mechanical zero: When a sighting system is adjusted to a completely centered state.

misfire: Failure of the propellant to ignite. In modern cartridges, this can be caused by a faulty primer or a light primer strike. In muskets, this can be caused by a clogged touch hole.

muzzle: The end of the barrel where the projectile exits at high velocity.

muzzle flash: The visible signature of a weapon being fired as flame and hot gases exit the barrel along with the projectile.

muzzleloader: Any firearm loaded through the muzzle.

needle gun: A firearm that used an elongated firing pin to pierce a paper cartridge and strike the primer contained within – thus firing the round.

negligent discharge: See accidental discharge.

nose: The rounded or pointed end of a bullet.

optics: A sighting system, scope, or holographic sight that uses formed lenses and an internal or projected aiming system.

over and under: Any firearm with one barrel mounted underneath the other barrel.

+P / +P+: Ammunition crafted to create higher pressures that will give the bullet a higher velocity. A firearm must be checked to see whether it can withstand the higher pressures generated by ammunition of this type.

pattern: The distribution of shot at various distances.

pepperbox gun: A firearm with multiple, single-shot barrels housed in a rotating mechanism.

pistol: A handheld firearm with one firing chamber. Pistols can be described by the sizes listed below.

pistol, full-size: Barrel length greater than 4.5 inches.

pistol, compact: Barrel length less than 4.5 inches, but greater than 3.5 inches.

pistol, subcompact: Barrel length less than 3.5 inches but greater than 3 inches.

pistol, pocket: A pistol with a barrel length less than 3 inches.

point blank: Extremely close range.

powder: The propellant used to fire a bullet.

powder, black: A less powerful propellant used to fire a bullet. Black powder is an original formulation of gun powder and produces more smoke and fouling than smokeless powder.

powder, smokeless: A modern formulation of propellant used to fire a bullet. Smokeless powder produces higher energy with a decreased amount of smoke and fouling.

press check: A way to see if a round is chambered. This is accomplished by retracting the bolt or slide far enough to visually confirm that the gun is loaded.

primer: A more energetic chemical compound used to ignite the main propellant.

printing: When a concealed firearm reveals itself as a bulge or shape under clothing.

propellant: A combustible or explosive mixture used to fire a bullet.

pull: See trigger pull.

rate of fire: A firearm's rate of fire can be broken down into two categories mechanical and operational. Usually expressed in rounds per minute (RPM), or rounds per second (RPS).

> **rate of fire, cyclic:** The mechanical maximum output the firearm can achieve. This rate ignores source of feed capacity and is strictly dependent on the construction of the gun.

> **rate of fire, rapid fire:** An operational rate of fire which temporarily ignores heat management in favor of superior fire power.

> **rate of fire, sustained:** An operational rate of fire which allows the firearm to be in constant use. This rate of fire considers heat management to prevent damage to barrel and other working parts.

rip – rack – roll – reload: An immediate action drill used to reduce the double feed stoppage.

receiver: The part of the rifle that houses all internal components of the firearm including, where applicable, the bolt, breech block, trigger assembly, and hammer. It also allows for the attachment of external components, such as the barrel, stock, and fore grip.

recoil: The backward force exerted on the firearm as the weapon is fired.

recoil pad: A device on the butt of a rifle or shotgun used to decrease experienced recoil.

reflex sight: See holo sight.

reload: (v.) Introducing ammunition into a gun that has expended all rounds through firing. (n.) Any ammunition that uses a casing that has been previously fired. These casings must first be cleaned and fitted with a new primer, propellant, and projectile.

revolver: A handheld pistol with multiple firing chambers contained in a cylinder.

rifle: A shoulder fired weapon with a rifled barrel.

rifling: A helical structure inside the barrel of a firearm made up of lands and grooves. Rifling imparts spin on the projectile, stabilizing its flight and increasing accuracy.

round: a common term used to describe both a bullet and a cartridge of ammunition.

safety: A mechanical device used to prevent the unintentional firing of a weapon.

sawed-off shotgun: A shotgun with a barrel that has been altered to a length below the legal limit.

scope: An optical aiming device which magnifies a distant target.

sear: A part of the firearm's trigger assembly that holds back the hammer or striker in a ready to fire position.

semi-automatic: See action, automatic, semi.

shell: A muti-use term used for shotgun ammunition both live and expended. It may also be used for expended rifle and pistol cartridges. See brass.

shotshell: Shotgun ammunition.

shot: A form of ammunition that uses multiple projectiles fired from a single barrel in one firing.

shot group: The pattern of bullet strikes fired at a single target during a single firing session. Three round groups are usually used when zeroing the sights or optics of a weapon. As a figure of speech, "Tighten up your shot group" is a polite way of saying, "Get your shit together."

shotgun: A smoothbore firearm designed to shoot pellets, shot, or slugs.

silencer: A misused term for suppressor.

skeet: A shooting sport in which the shooter attempts to hit a clay target while it's in flight.

slide: An assembly of parts in a semi-automatic pistol that performs the same action as the bolt in a rifle.

slow match: A chemically treated cord designed to burn slowly and consistently. This was the primary ignition source in the first muskets.

slug: A type of shotgun ammunition where, instead of multiple projectiles, there is only one.

snub nose: A handgun, usually a revolver, with an extremely short barrel.

sporting clay: A flying target used in skeet shooting.

S.P.O.R.T.S: An immediate action drill used to reduce stoppages in AR variant rifles.

standoff: The distance between the gun's sights and the center of the bore of the barrel.

stock: The part of the longarm that rests against the shoulder. In muskets, the stock acts as a receiver in modern firearms.

stove pipe: A type of stoppage or jam that occurs when a spent cartridge is improperly ejected.

stoppage: A term used to describe an unintentional break in the firing cycle. Stoppages are caused by faulty ammunition, faulty magazines, or in the case of semi-automatic pistols, poor weapon control.

striker: See firing pin.

submachine gun: Any firearm designed to fire in the fully automatic mode that uses handgun ammunition.

suppressor: A device attached to the barrel of a firearm used to reduce the amount of expanding gas leaving the barrel when a round is fired.

tactical reload: the act of inserting more ammunition into a gun's magazine while in an engagement or battle. For tube magazines this means simply inserting more cartridges. For ejecting box magazines, it means removing the partially fired magazine and inserting a full one.

tap – rack – bang: An immediate action drill to reduce stoppages in semi-automatic pistols.

thumbing: A quick draw technique for single action revolvers where the shooter draws back the hammer with the thumb of the firing hand while drawing the revolver from its holster.

touch hole: In muskets, a small hole through which flame travels from the pan to the powder in the barrel.

trajectory: The flight path of a projectile from muzzle to target.

trapdoor: An early design for breech-loading rifles. Also, some firearms with internal box magazines have a trapdoor that allows for easy unloading without having to cycle the action until the magazine is clear.

trigger: The part of a firearm that a shooter activates to fire a round.

trigger, hair: A trigger which requires less force to activate.

trigger lock: A form of safety.

trigger pull: The amount of force required, measured in pounds, to activate a trigger.

velocity: The speed at which a bullet travels.

wad/wadding: Material used to insulate a projectile or shot during the firing of a gun.

wheellock: A firing system for muskets that used a spring-driven wheel to create sparks that would ignite the primer charge in the pan.

Winchester: A term used by the US military that means you've expended all your ammunition.

zero: The sum of adjustments needed to make point of aim equal to point of impact at a given distance.

zeroing: The procedure used to make the point of aim the same as the point of impact at a specific distance. This is performed in incremental steps where the strike of the rounds fired are observed and adjustments are made to the sighting system.

zip gun: A crude firearm made with unrifled pipe for a barrel and a rudimentary firing system.

Bibliography

Sorted by major category (Note: some references were used for multiple topics.)

Strunk, William and E.B. White. *The Elements of Style*. Needham Heights, MA: Pearson, 2000

Muzzle-Loaders

Bilby, Joseph G. *Civil War Firearms Their Historical Background, Tactical Use and Modern Collecting and Shooting*. Buchanan, NY: Combined Books, 1996.

"Meet a Musketeer." Weapons and Warfare. British Civil Wars. Accessed April 14, 2024. https://britishcivilwars.ncl.ac.uk/weapons-warfare/meet-the-civil-war-soldiers/meet-a-musketeer/

Hogg, Ian V. *Guns and How They Work*. Edited by L.K. Wood. London: Marshall Cavendish, 1979.

Holcombe, Colin. *A History of Firearms*. Self-published, Amazon Digital Services, 2018.

"Rifled Musket." Military Wiki. Accessed April 14, 2024. https://military-history.fandom.com/wiki/Rifled_musket

Stebbins, Samuel. "Every Standard Issue US Military Rifle Since the American Revolution." 24/7 Wall ST July 4, 2023. https://247wallst.com/special-report/2023/07/04/every-standard-issue-us-military-rifle-since-the-american-revolution/#:~:text=The%20U.S.%20Model%201842%20musket,parts%2C%20making%20for%20easier%20repairs

Matchlock

History Hit. "How to Fire a Matchlock Musket" https://www.youtube.com/watch?v=2KTS8PQ06Qo

"Matchlock." Wikipedia. Last edited March 20, 2024. https://en.wikipedia.org/wiki/Matchlock

Wheellock

JFY Museum. "Primed and Loaded | Wheelock Ignition System"
    https://www.youtube.com/watch?v=9YRg2fhy19Q

"Wheellock." Wikipedia. Last edited February 3, 2024.
    https://en.wikipedia.org/wiki/Wheellock

Flintlock

Reid, Stuart. *The Flintlock Musket: Brown Bess and Charleville 1715–1865 (Weapon Book 44)*. Oxford: Osprey, 2016. Kindle.

"Flintlock." Wikipedia. Last edited April 8, 2024.
    https://en.wikipedia.org/wiki/Flintlock

Caplock

"Alexander John Forsyth." Wikipedia. Last edited May 30, 2023.
    https://en.wikipedia.org/wiki/Alexander_John_Forsyth#cite_note-Houze_1991_p._37-10

Bocetta, Sam. "The Complete History of Small Arms Ammunition and Cartridges." Small Wars Journal October 15, 2017.
    https://247wallst.com/special-report/2023/07/04/every-standard-issue-us-military-rifle-since-the-american-revolution/#:~:text=The%20U.S.%20Model%201842%20musket,parts%2C%20making%20for%20easier%20repairs

"Percussion Cap." Wikipedia. Last edited February 3, 2024.
    https://en.wikipedia.org/wiki/Percussion_cap

Smithurst, Peter. *The Pattern 1853 Enfield Rifle (Weapon Book 10)*. Oxford: Osprey, 2011. Kindle.

Loading

"Bandolier." Wikipedia. Last edited April 13, 2024.
    https://en.wikipedia.org/wiki/Bandolier

Black Powder Maniac Shooter. "Running while Reloading a Flintlock Rifle" https://www.youtube.com/watch?v=bflPncephRQ

Croz40. "1764 Manual of Arms Firing Procedure"
    https://www.youtube.com/watch?v=SuYGCji-_5A

LionHeart FilmWorks. "Civil War - Musket Loading Drill 'In-Nine-Times' HD"
    https://www.youtube.com/watch?v=VCAYXQ1Z6q4&t=29s

"Paper cartridge." Wikipedia. Last edited December 25, 2023.
https://en.wikipedia.org/wiki/Paper_cartridge

Valis, Glenn, transcriber. "The Manual Exercise, As ordered by his Majesty, In 1764." Double GV. Accessed April 14, 2024, http://www.doublegv.com/ggv/battles/Manual.html

Blunderbuss
"Blunderbuss." Wikipedia. Last edited April 8, 2024.
https://en.wikipedia.org/wiki/Blunderbuss

InRange TV. "Blunderbuss - Buckshot Patterning"
https://www.youtube.com/watch?v=gWbJPI6sNCU

Breech-Loaders
Conversions
Canfield, Bruce N. "Origins Of The 'Trapdoor' Springfield: The Allin Conversions." American Rifleman. August 7, 2020.
https://www.americanrifleman.org/content/origins-of-the-trapdoor-springfield-the-allin-conversions/

"M1854/67 & M1862/67 Austrian Wänzl." Military Rifles. Updated August 18, 2022. https://www.militaryrifles.com/austria/wanzl

TFB TV. "Springfield Trapdoor: America's Breech-Loader"
https://www.youtube.com/watch?v=JC0C41KX8RY

Wilkinson, Fred, "The Adoption of the Snider, and Its Use in an Early Battle." Articles, Manuals & Catalogs. Historical Breechloading Smallarms Association. Accessed April 14, 2024. https://hbsa-uk.org/knowledge-and-research/articles/snider-conversion-and-magdala/

Dreyes Needle Gun
Bilby, Joseph G. *A Revolution in Arms: A History of the First Repeating Rifles (Weapons in history)*. Westholme Publishing. Kindle Edition.

"Dreyse needle gun." Wikipedia. Last edited April 25, 2024.
https://en.wikipedia.org/wiki/Dreyse_needle_gun

"Needle gun." Wikipedia. Last edited May 3, 2024.
https://en.wikipedia.org/wiki/Needle_gun

McCollum, Ian. "The Last Dreyse Needlefire: 1874 Border Guard."
Forgotten Weapons March 13, 2019.
https://www.youtube.com/watch?v=5qL1g8Hjcpk

Németh, Balázs. *Early Military Rifles: 1740–1850 (Weapon Book 76).*
Bloomsbury Publishing. Kindle Edition.

Hall Rifle

McCollum, Ian. "Hall Model 1819: A Rifle to Change the Industrial
World." Forgotten Weapons September 7, 2020.
https://www.youtube.com/watch?v=vpW054cVfHc

"M1819 Hall rifle." Wikipedia. Last editedMarch 26, 2024.
https://en.wikipedia.org/wiki/M1819_Hall_rifle

Stalvo, David. "Shooting the 1843 Hall Carbine." August 30, 2023.
https://www.youtube.com/watch?v=B0IP0Dq3w1o

Jean Samuel Pauly

"Jean Samuel Pauly." Wikipedia. Last edited April 10, 2024.
https://en.wikipedia.org/wiki/Jean_Samuel_Pauly

McCollum, Ian. "Samuel Pauly Invents the Cartridge in 1812." Forgotten
Weapons May 4, 2019.
https://www.youtube.com/watch?v=lHuNo2XU57g

Other Actions

3DGunner. "3D Animation: How a Shotgun Works"
https://www.youtube.com/watch?v=ZbV3jkgaEek

"Actions: Break Open Action." Firearms History, Technology &
Development. July 5, 2010.
http://firearmshistory.blogspot.com/2010/07/actions-break-open-
action.html

BC's Gun Channel. "Winchester/Browning Model 1885 High Wall Rifle
.45-70 Gov't" https://www.youtube.com/watch?v=SJFnDc_m5SY

Layman, George. "The Remington Rolling Block." HistoryNet. January
1, 2020. https://www.historynet.com/the-Remington-rolling-block/

TFB TV. " The Remington Rolling Block Rifle."
https://www.youtube.com/watch?v=FNv8hpxDDkU

Repeating Rifles
Pump-Action Longarms

McCollum, Ian. "Testing Slamfire: Sneaky Advantage or Useless Hype?" Forgotten Weapons September 23, 2023.
https://www.youtube.com/watch?v=jksldX33HAY

"Ithaca 37." Wikipedia. Last edited January 28, 2024.
https://en.wikipedia.org/wiki/Ithaca_37

"Mossberg 500." Wikipedia. Last edited February 10, 2024.
https://en.wikipedia.org/wiki/Mossberg_500

Rittman, Matt. "How a Pump Shotgun Works"
https://www.youtube.com/watch?v=21uh28Z77Xg

"Remington Model 7600." Wikipedia. Last edited March 13, 2024.
https://en.wikipedia.org/wiki/Remington_Model_7600

Thompson, Leroy. *US Combat Shotguns (Weapon Book 29)*. Oxford: Osprey, 2013. Kindle.

Lever-Action Rifles

3DGunner. "3D Animation: How a Lever Action Rifle Works (Marlin)"
https://www.youtube.com/watch?v=58LbxVd4buo

Duelist1954. "Armi Sport 1865 Spencer Carbine.mov"
https://www.youtube.com/watch?v=WwhLuhRWYyI

Hickok45. "1860 Henry Rifle"
https://www.youtube.com/watch?v=NbfXjqDzago

"Lever action." Wikipedia. Last edited April 13, 2024.
https://en.wikipedia.org/wiki/Lever_action

Paper Cartridges. "The 13 Tube Blakeslee Box for the Spencer Repeating Civil War Carbine"
https://www.youtube.com/watch?v=v4EIB_TiSUk

Pegler, Martin. *Winchester Lever-Action Rifles (Weapon Book 42)*. Oxford: Osprey, 2015. Kindle.

Bolt-Action Rifles

Best Sniper Simulator. "How Bolt Action Rifle Works (Remington 700 Mechanism) https://www.youtube.com/watch?v=cGbV8hp0pqU

"Bolt action." Wikipedia. Last edited November 5, 2023 https://en.wikipedia.org/wiki/Bolt_action

Paper Cartridges. "A Civil War Bolt Action! Paper Cartridges Shoots the Greene Rifle" https://www.youtube.com/watch?v=hl4fPfDWiDk

Sootch00. "Lee Enfield No.1 Mk III SMLE Rifle Review : 'Smelly'" https://www.youtube.com/watch?v=nSJHmLVAwNI

Sootch00. "M1903A3 Springfield Rifle Review" https://www.youtube.com/watch?v=yJjKH7nPJas

Thompson, Leroy. *The M1903 Springfield Rifle (Weapon Book 23)*. Oxford: Osprey, 2013. Kindle.

Automatic Rifles

McCollum, Ian. "Mannlicher 1885 Semiauto Rifle." Forgotten Weapons. May 6, 2015. https://www.forgotenweapons.com/mannlicher-1885-semiauto-rifle/

M1 Garand

Garand Thumb. "I Give Myself Garand Thumb/We Find Out How Painful Garand Thumb Is" https://www.youtube.com/watch?v=ssvZWdu4sZ0

"M1 Garand." Wikipedia. Last edited April 12, 2024. https://en.wikipedia.org/wiki/M1_Garand

Thompson, Leroy. *The M1 Garand (Weapon Book 16)*. Oxford: Osprey, 2012. Kindle.

FN FAL

Cashner, Bob. *The FN FAL Battle Rifle (Weapon Book 27)*. Oxford: Osprey, 2013. Kindle.

"FN Model 1910." Wikipedia. Last edited April 14, 2024. https://en.wikipedia.org/wiki/FN_Model_1910

AK47

"AK-47." Wikipedia. Last edited March 27, 2024.
https://en.wikipedia.org/wiki/AK-47

Rottman, Gordon L. *The AK-47: Kalashnikov-Series Assault Rifles (Weapon Book 8)*. Oxford: Osprey, 2011. Kindle.

M14

"M14 rifle." Wikipedia. Last edited April 14, 2024.
https://en.wikipedia.org/wiki/M14_rifle

Thompson, Leroy. *The M14 Battle Rifle (Weapon Book 37)*. Oxford: Osprey, 2014. Kindle.

M16

"M16 rifle." Wikipedia. Last edited April 8, 2024.
https://en.wikipedia.org/wiki/M16_rifle

Rottman, Gordon L. *The M16 (Weapon Book 14)*. Oxford: Osprey, 2011. Kindle.

Steyr Aug

3DGunner. "3D Animation: How a Steyr AUG Bullpup Rifle Works"
https://www.youtube.com/watch?v=gIOUv7aFbto

"Steyr AUG." Wikipedia. Last edited March 24, 2024.
https://en.wikipedia.org/wiki/Steyr_AUG

Machine Guns

Watson, Stephanie and Tom Harris. "How the Machine Gun Revolutionized Warfare." How Stuff Works. Last updated September 12, 2023. https://science.howstuffworks.com/machine-gun.htm

Gatling Gun

"Gatling gun." Wikipedia. Last edited April 4, 2024.
https://en.wikipedia.org/wiki/Gatling_gun

"M134 Minigun." Wikipedia. Last edited March 27, 2024.
https://en.wikipedia.org/wiki/M134_Minigun

"M61 Vulcan." Wikipedia. Last edited April 2, 2024.
https://en.wikipedia.org/wiki/M61_Vulcan

Smithurst, Peter. *The Gatling Gun (Weapon Book 40)*. Oxford, Osprey, 1015. Kindle.

Lewis Gun

"Lewis gun." Wikipedia. Last edited April 4, 2024.
   https://en.wikipedia.org/wiki/Lewis_gun

Maxim Gun

Forces News. "World's Longest-Serving Machine Gun Still Used on Ukraine Frontline"
   https://www.youtube.com/watch?v=nBrCChDxbKw

John H. Lienhard. "Hiram Maxim" in The Engines of Our Ingenuity. No. 694. Cullen College of Engineering. University of Houston.
   https://engines.egr.uh.edu/episode/694.

"Maxim gun." Wikipedia. Last edited March 29, 2024.
   https://en.wikipedia.org/wiki/Maxim_gun

Pegler, Martin. *The Vickers-Maxim Machine Gun (Weapon Book 25)*. Oxford: Osprey, 2013. Kindle.

MG 34

Forgotten Weapons. "Shooting the MG-34 and MG-42"
   https://www.youtube.com/watch?v=GfJkU4Sah8I

McNab, Chris. *MG 34 and MG 42 Machine Guns (Weapon Book 21)*. Oxford: Osprey, 1012. Kindle.

"MG 34." Wikipedia. Last edited March 7, 2024.
   https://en.wikipedia.org/wiki/MG_34

M60 Machine Gun

Dockery, Kevin. *The M60 Machine Gun (Weapon Book 20)*. Oxford: Osprey, 2012. Kindle.

Forgotten Weapons. "Original Vietnam-Era M60 at the Range"
   https://www.youtube.com/watch?v=uJ2LJVcuoaM

Marine Mindset. "US Marines Demonstrating 'Talking Guns' with Soviet PKM Machine Guns"
   https://www.youtube.com/watch?v=ThsGaj6_Cpg

"M60 machine gun." Wikipedia. Last edited March 17, 2024.
   https://en.wikipedia.org/wiki/M60_machine_gun

Submachine Guns

"Submachine gun." Wikipedia. Last edited March 26, 2024,
   https://en.wikipedia.org/wiki/Submachine_gun

MP5

Forgotten Weapons. "At the Range with the Iconic MP5A3"
   https://www.youtube.com/watch?v=vkzelAfhsvA

"Heckler & Koch MP5." Wikipedia. Last edited March 9, 2024.
   https://en.wikipedia.org/wiki/Heckler_%26_Koch_MP5

Thompson, Leroy. *The MP5 Submachine Gun (Weapon Book 35)*. Oxford:
   Osprey, 2014. Kindle.

Thompson

Pegler, Martin. *The Thompson Submachine Gun: From Prohibition Chicago to
   World War II (Weapon Book 1)*. Oxford: Osprey, 2011. Kindle.

TFB TV. "The Thompson M1A1 Submachine Gun (Full Auto)"
   https://www.youtube.com/watch?v=D53opaeollQ&t=216s

"Thompson submachine gun." Wikipedia. Last edited April 15, 2024.
   https://en.wikipedia.org/wiki/Thompson_submachine_gun

Uzi

McNab, Chris. *The Uzi Submachine Gun (Weapon Book 12)*. Oxford:
   Osprey, 2011. Kindle.

TFB TV. "The Uzi Submachine Gun (Full Auto)"
   https://www.youtube.com/watch?v=ayZClbRuLTQ

"Uzi." Wikipedia. Last edited March 20, 2024.
   https://en.wikipedia.org/wiki/Uzi

Derringers

Sootch. "Bond Arms 410 & 45LC Derringer"
https://www.youtube.com/watch?v=5SoWsLdxUTs

.357 COP

"COP .357 Derringer." Wikipedia. Last updated September 11, 2023.
https://en.wikipedia.org/wiki/COP_.357_Derringer

Forgotten Weapons. "The Most 80s Gun Ever: COP 357 at the Backup
Gun Match" https://www.youtube.com/watch?v=P4ULyMeh3bw

TTAG Contributor. "Gun Review: COP .357 Derringer Pistol." The
Truth About Guns. July 9, 2018.
https://www.thetruthaboutguns.com/gun-review-cop-357-derringer-
pistol/

Philadelphia Derringer

"Derringer." Wikipedia. Last updated December 31, 2023.
https://en.wikipedia.org/wiki/Derringer

Martin, Craig. "History of the Derringer: Concealed Carry's Forefather."
Concealed Carry. January 11, 2018.
https://www.concealedcarry.com/firearm-history/history-derringer-
concealed-carrys-forefather/

Remington Model 95 Derringer

Blanchard, Steve. "Shooting an Original .41 Rimfire Remington Double
Derringer." https://www.youtube.com/watch?v=XeQVLCbdXGk

"Remington Model 95 Double Derringer – Exceptional." College Hill
Arsenal. Accessed April 14, 2024.
https://collegehillarsenal.com/remington-model-95-double-derringer-
exceptional

"Remington Model 95." Wikipedia. Last updated March 31, 2024.
https://en.wikipedia.org/wiki/Remington_Model_95

Sharps Derringer

Fattywithafirearm. "1861 C. Sharps 4 Shot Derringer"
https://www.youtube.com/watch?v=5k--QgN38HY

"Sharps Model 1C Pepperbox. College Hill Arsenal. Accessed April 14,
2024. https://collegehillarsenal.com/sharps-model-1c-pepperbox

Revolvers

Cisko Master Gunfighter. "How To Shoot Fast Draw – Thumbing & Fanning" https://www.youtube.com/watch?v=wr0azVhRZUQ

Gun Sam_Revolver Aficionado_. "How To SAFELY De-Cock a Revolver" https://www.youtube.com/watch?v=O8-cbgzeQSY

"Smith & Wesson Model 40 Rare Flat Latch Grip Safety 38 Spl Mfg 1955-1957." Guns International. Accessed April 14, 2024. https://www.gunsinternational.com/guns-for-sale-online/revolvers/smith---wesson-revolvers/smith---wesson-model-40-rare-flat-latch-grip-safety-38-spl-mfg-1955-1957.cfm?gun_id=101081856

Allen and Thurber Pepperbox

"Allen & Thurber Pepperbox Revolver Worchester Mass 49ers Gold Rush Antique." Ancestry Guns. Accessed April 14, 2024. https://www.ancestryguns.com/shop/allen-thurber-pepperbox-revolver-worchester-mass-49ers-gold-rush-antique-6-shot-32-revolver-from-the-1840s/

InRangeTV. "1850's Self Defense: The Allen & Thurber Pepperbox" https://www.youtube.com/watch?v=wjReSGFtUtY

"Pepper-box." Wikipedia. Last edited April 13, 2024. https://en.wikipedia.org/wiki/Pepper-box

Collier Flintlock Revolver

3D Gunner. "3D Animation: How the Collier Flintlock Revolver Worked" https://www.youtube.com/watch?v=Pfm06EcBtcc

"Elisha Collier." Wikipedia. Last edited October 19, 2023. https://en.wikipedia.org/wiki/Elisha_Collier

Forgotten Weapons. "Collier Flintlock Revolvers" https://www.youtube.com/watch?v=i9Km5KaeO7I

Colt Peacemaker

"Colt Single Action Army." Wikipedia. Last edited March 2, 2024, https://en.wikipedia.org/wiki/Colt_Single_Action_Army

Krakower, Will. "Colt Peacemaker: How the Colt Single Action Army Won the West." Oldwest. Last updated October 20, 2023. https://www.oldwest.org/colt-peacemaker-single-action-army/

TFB TV. "Colt Single Action Army: Shooting The Legendary Revolver" https://www.youtube.com/watch?v=sVbTnpzByjY

Lefaucheux M1854

Curiosity Incorporated. "1800's LeFaucheux Pin Fire Revolver, Walk Through" https://www.youtube.com/watch?v=yN80jIz4Ebo

"Fine M1854 Lefaucheux Revolver." College Hill Arsenal. Accessed April 14, 2024. https://collegehillarsenal.com/fine-m1854-lefaucheux-revolver

"Lefaucheux M1854." Wikipedia. Last edited January 14, 2024. https://en.wikipedia.org/wiki/Lefaucheux_M1854

Navy Colt

"Colt 1851 Navy Revolver." Wikipedia. Last edited April 2, 2024. https://en.wikipedia.org/wiki/Colt_1851_Navy_Revolver

"Colt 1851." OutlawsColts. Accessed April 14, 2024. https://outlawscolts.jouwweb.nl/colt-1851

"Colt Model 1851 Navy Revolver." National Museum of American History. Accessed April 14, 2024. https://americanhistory.si.edu/collections/nmah_1005116

Frontier Western Heritage. "Shooting the 1851 Navy Colt. https://www.youtube.com/watch?v=2L30-McaZCA

Pfeifer Zeliska .600 Nitro Express Revolver

Lewis, Vince. "Pfeifer-Zeliska .600 Nitro Express Handgun: 'How much gun is TOO much gun?!'" Tony Rogers. Last updated December 4, 2009. https://www.tonyrogers.com/weapons/pfeifer-zeliska.htm

"Pfeifer Zeliska .600 Nitro Express revolver." Wikipedia. Last edited March 5, 2024. https://en.wikipedia.org/wiki/Pfeifer_Zeliska_.600_Nitro_Express_revolver

Reichs Revolver

Forgotten Weapons. "Shooting the 1883 Reichsrevolver" https://www.youtube.com/watch?v=RB074qA-3EA

"M1879 Reichsrevolver." Wikipedia. Last edited October 1, 2023 https://en.wikipedia.org/wiki/M1879_Reichsrevolver

Shell, Bob. "The Model 1883 Reichsrevolver." Guns and Ammo. October 22, 2018. https://www.handgunsmag.com/editorial/the-model-1883-reichsrevolver/326486

Smith & Wesson Safety Hammerless

MidwayUSA. "Smith and Wesson 32 Safety Hammerless First Model"
  https://www.youtube.com/watch?v=fbIC84MVK5s

"The Safety Hammerless Revolvers: A Brief History." The Smith & Wesson Forum. Accessed April 14, 2024. https://smith-wessonforum.com/s-w-antiques/283209-safety-hammerless-revolvers-brief-history.html

"Smith & Wesson 32 Safety First Model revolver." NRA Museums. Accessed April 14, 2024. https://www.nramuseum.org/guns/the-galleries/innovation,-oddities-and-competition/case-26-the-booming-arms-industry/smith-wesson-32-safety-first-model-revolver.aspx

Tranter

"Tranter Patent M-1863 .450CF Revolver." College Hill Arsenal. Accessed April 14, 2024. https://collegehillarsenal.com/tranter-patent-m-1863-450cf-revolver

"Tranter (revolver)." Wikipedia. Last edited June 7, 2023. https://en.wikipedia.org/wiki/Tranter_(revolver)

Webley-Fosbery

Forgotten Weapons. "Shooting the Webley-Fosbery Automatic Revolver - Including Safety PSA"
  https://www.youtube.com/watch?v=4EqkcVlzVSw

"Webley-Fosbery Automatic Revolver." National Museum of American History. Accessed April 14, 2024.
  https://americanhistory.si.edu/collections/nmah_415916

"Webley–Fosbery Automatic Revolver." Wikipedia. Last edited November 21, 2023.
  https://en.wikipedia.org/wiki/Webley%E2%80%93Fosbery_Automatic_Revolver

Webley Mark 1

Skallagrim. "A Hands-On Look at the Old Webley Revolver"
  https://www.youtube.com/watch?v=yxinQQx0_ZA

"Webley MK 1 Revolver." NRA Museums. Accessed April 14, 2024.
  https://www.nramuseum.org/guns/the-galleries/world-war-i-and-firearms-innovation/case-32-wwi-america-and-the-allies/webley-mk-i-revolver.aspx

"Webley Revolver." Wikipedia. Last edited April 15, 2024.
https://en.wikipedia.org/wiki/Webley_Revolver

Semi-Automatic Pistols

Baker, Chris. "The Rise and Fall of the Double Action Semi-Auto." Lucky Gunner. May 5, 2016.
https://www.luckygunner.com/lounge/rise-fall-double-action-semi-auto/

PreppingDad. "Shadow Systems DR920P Explosion"
https://www.youtube.com/watch?v=gnVf62hOt-g

Borchardt C93

"Borchardt Automatic Pistol." Forgotten Weapons. Accessed April 14, 2024. https://www.forgottenweapons.com/early-automatic-pistols/borchardt-automatic-pistol/

"Borchardt C-93." Wikipedia. Last edited March 26, 2024.
https://en.wikipedia.org/wiki/Borchardt_C-93

Forgotten Weapons. "C93 Borchardt: the First Successful Self-Loading Pistol" https://www.youtube.com/watch?v=ItpOBQFVIhM

Beretta 92f

"Beretta M9." Wikipedia. Last edited February 9, 2024.
https://en.wikipedia.org/wiki/Beretta_M9

Sootch00. "Beretta 92F or M9 9mm Service Pistol"
https://www.youtube.com/watch?v=MWZq2E8MXw8

Thompson, Leroy. *The Beretta M9 Pistol (Weapon Book 11).* Oxford: Osprey, 2011. Kindle.

FN Model 1910

"FN Model 1903." Wikipedia. Last edited February 21, 2024
https://en.wikipedia.org/wiki/FN_Model_1903

"FN Model 1910." Wikipedia. Last edited April 14, 2024.
https://en.wikipedia.org/wiki/FN_Model_1910

Sootch00. "FN Model 1910 32 ACP Gun Review"
https://www.youtube.com/watch?v=OQaDq2n1hTo

Grendel p10

"Before There Was Kel Tec: The Evolutionary Grendel P10." Last Stand on Zombie Island. March 13, 2015. https://laststandonzombieisland.com/2015/03/13/before-there-was-kel-tec-the-evolutionary-grendel-p10/

"Grendel Inc." Wikipedia. Last edited February 28, 2024. https://en.wikipedia.org/wiki/Grendel_Inc.

GunWebsites. "No Magazine Semi-Auto Pistol - Grendel P-10" https://www.youtube.com/watch?v=3F4QRP4MH5I&t=85s

Luger P08

Grant, Neil. *The Luger (Weapon Book 64)*. Oxford: Osprey, 2018. Kindle.

"Luger pistol." Wikipedia. Last edited April 8, 2024. https://en.wikipedia.org/wiki/Luger_pistol

Sootch00. "The German Luger P08 Gun Review" https://www.youtube.com/watch?v=uH4yxffwhqA

M1911

"M1911 pistol." Wikipedia. Last edited April 13, 2024. https://en.wikipedia.org/wiki/M1911_pistol

Student of the Gun. "M1911 A1 – Saigon Report Ep. 01" https://www.youtube.com/watch?v=fUzyf-Ik9m4

Thompson, Leroy. *The Colt 1911 Pistol (Weapon Book 9)*. Oxford: Osprey, 2011. Kindle.

Mauser C96

Ferguson, Jonathan. *The 'Broomhandle' Mauser (Weapon Book 58)*. Oxford: Osprey, 2017. Kindle.

"Mauser C96." Wikipedia. Last edited April 13, 2024. https://en.wikipedia.org/wiki/Mauser_C96

Nambu Pistols

Military Arms Channel. "WWII Japanese Type 14 Nambu" https://www.youtube.com/watch?v=pk2XMWQigu4&t=612s

"Nambu pistol." Wikipedia. Last edited January 31, 2024. https://en.wikipedia.org/wiki/Nambu_pistol

Walter, John. *Nambu Pistols: Japanese military handguns 1900–45 (Weapon Book 86)*. Oxford: Osprey, 2023. Kindle.

Rohrbaugh R9

"Rohrbaugh R9." Wikipedia. Last edited August 3, 2022.
https://en.wikipedia.org/wiki/Rohrbaugh_R9

TheYankeeMarshal. "Rohrbaugh R9s: Range Report"
https://www.youtube.com/watch?v=uvf4KI3OX0c&t=350s

Ruger Mk IV

Military Arms Channel. "New Ruger Mk IV 22/45 Tactical with DeadAir
MASK 22!" https://www.youtube.com/watch?v=oWI0yydiyzc

"Ruger Standard." Wikipedia. Last edited January 25, 2024.
https://en.wikipedia.org/wiki/Ruger_Standard

Salvator-Dormus Pistol

"1891 Salvatore-Dormus: The First Automatic Pistol." Forgotten
Weapons. May 15, 2017. https://www.forgottenweapons.com/1891-
salvatore-dormus-the-first-automatic-pistol/

Military Arms Channel. "Affordable Combat Collectibles - Mauser C96
Broom Handle"
https://www.youtube.com/watch?v=QHFuOAhB_lA&t=640s

"Salvator-Dormus 1891." Gun Wiki. Accessed April 14, 2024.
https://guns.fandom.com/wiki/Salvator-Dormus_1891

"Salvator Dormus pistol." Wikipedia. Last edited March 19, 2023.
https://en.wikipedia.org/wiki/Salvator_Dormus_pistol

Schönberger-Laumann, 1892

Forgotten Weapons. "Laumann 1891 and Schonberger-Laumann 1894
Semiauto Pistols"
https://www.youtube.com/watch?v=AFhU3Dixvnk&t=538s

"Laumann 1891 and Schonberger-Laumann 1894 Semiauto Pistols."
Forgotten Weapons. May 22, 2017.
https://www.forgottenweapons.com/laumann-1891-and-
schonberger-laumann-1894-semiauto-pistols/

"Schönberger-Laumann 1892." Wikipedia. Last edited April 11, 2024.
https://en.wikipedia.org/wiki/Schönberger-Laumann_1892

Shadow Systems CR920

"The Covert 920." Shadow Systems Corp. Accessed April 14, 2024.
https://shadowsystemscorp.com/cr920/

Honest Outlaw. "Shadow Systems CR920 1000 Round Review: When Expectation Meets Reality"
https://www.youtube.com/watch?v=NKbF35iNMU0

Sig Sauer P226

Military Arms Channel. "Sig Sauer P226 9mm"
https://www.youtube.com/watch?v=aHI2B6Qlkg4

"SIG Sauer P226." Wikipedia. Last edited April 6, 2024.
https://en.wikipedia.org/wiki/SIG_Sauer_P226

Walther PP

Campbell, Dave. "A Look Back at the Walther PP." American Rifleman. April 6, 2018. https://www.americanrifleman.org/content/a-look-back-at-the-walther-pp/

Military Arms Channel. "WWII era Walther PP Pistol"
https://www.youtube.com/watch?v=tx5IKyu0wUA

"Walther PP." Wikipedia. Last edited March 23, 2024.
https://en.wikipedia.org/wiki/Walther_PP

Machine Pistols
Glock 18

"Glock" Wikipedia. Last edited April 10, 2024.
https://en.wikipedia.org/wiki/Glock#Glock_18

TFB TV. "The Full Auto Glock 18C Machine Pistol"
https://www.youtube.com/watch?v=cc4ZTlsusTQ

Ingram M10

Forgotten Weapons. "Ingram M10 & M11 SMGs: The Originals from Powder Springs"
https://www.youtube.com/watch?v=BU9kVinACYU

Iannamico, Frank. "MAC 10." Small Arms Review. March 23, 2021.
https://smallarmsreview.com/mac-10/

"MAC-10." Wikipedia. Last edited April 7, 2024.
https://en.wikipedia.org/wiki/MAC-10

Škorpion VZ 61

Forgotten Weapons. "Shooting the Czech vz61 Škorpion: Machine Pistol or PDW?" https://www.youtube.com/watch?v=TED6z4_9fmw

Guttman, Jon. "Czechoslovakia Delivered a Sting With Its Škorpion VZ 61 Subgun (Or Machine Pistol?)." Military Times. March, 5, 2020. https://www.militarytimes.com/off-duty/gearscout/tacticool/2020/03/05/czechoslovakia-delivered-a-sting-with-its-skorpion-vz-61-subgun-or-machine-pistol/

"Škorpion." Wikipedia. Last edited March 29, 2024. https://en.wikipedia.org/wiki/Škorpion

Ammunition

"A Guide to Understanding Cartridge Names." Gun Tweaks. Accessed April 14, 2024. https://www.guntweaks.com/understanding-cartridge-names.html

"Ammunition." Wikipedia. Last edited April 9, 2024. https://en.wikipedia.org/wiki/Ammunition

"Ballistics," Wikipedia. Last edited July 29, 2023. https://en.wikipedia.org/wiki/Ballistics

"Calibers." Pistol Basics 101. Accessed April 14, 2024. https://www.pistolbasics101.com/calibers--bullets-101.html

"Cartridge (firearms)." Wikipedia. Last edited April 11. 2024. https://en.wikipedia.org/wiki/Cartridge_(firearms)

"Caseless Ammunition." Wikipedia. Last edited March 21. 2024. https://en.wikipedia.org/wiki/Caseless_ammunition

"Crime Data: Law Enforcement Officers Killed in the Line of Duty Statistics for 2021." Law Enforcement Bulletin. FBI. November 9, 2022. https://leb.fbi.gov/bulletin-highlights/additional-highlights/crime-data-law-enforcement-officers-killed-in-the-line-of-duty-statistics-for-2021

FR Author Group. "Color-Coded Bullets Tips [Tables]: Different Types and Their Uses." Forensic Reader Accessed April 14, 2024. https://forensicreader.com/color-coded-bullets-tables/?expand_article=1

"Headspace (firearms)." Wikipedia. Last edited February 4, 2024.
https://en.wikipedia.org/wiki/Headspace_(firearms)

"Infrared Tracer." CBD Defense. Accessed April 14, 2024.
https://cbcdefense.com/special-technologies/infrared-tracer/

Jones, Allan. "How Cartridges Get Their Names." Shooting Times.
February 27, 2018. https://www.shootingtimes.com/editorial/how-cartridges-get-their-names/99501

Justice Technology Information Center. "Understanding NIJ 0101.06
Armor Protection Levels." US Department of Justice Office of Justice
Programs. Accessed April 20, 2024.
https://www.ojp.gov/pdffiles1/nij/nlectc/250144.pdf.

"Pinfire cartridge." Wikipedia. Last edited February 29, 2024.
https://en.wikipedia.org/wiki/Pinfire_cartridge

Mayer, Scott E. "Banging Out Boattails." RifleShooter. January 4, 2011.
https://www.rifleshootermag.com/editorial/ammunition_rs_boattails_093009wo/84178.

"Shot Size Chart and Guide for Upland Birds, Waterfowl, and Turkeys."
Field and Stream. Updated March 3, 2203.
https://www.fieldandstream.com/photos/gallery/guns/shotguns/ammunition/2011/07/fss-ultimate-shotshell-guide/

"Shotgun Shells – a High Level Overview." Hiking, Camping and
Shooting. November 11, 2017.
https://www.hikingcampingandshooting.com/blog/shotgun-shells-a-high-level-overview

Willis, Andrew. "Why Handgun and Rifle Bullets are Different." Chuck
Hawks. Accessed April 14, 2024.
https://chuckhawks.com/handgun_rifle_bullets.htm#

"Wound Ballistics 2-Mechanics of Projectile Wounding." Viper Blog.
Viper Weapons. August 18, 2023.
https://viperweapons.us/blog/f/mechanics-of-projectile-wounding

Historical Events
Miami Shootout

"1986 FBI Miami shootout." Wikipedia. Last edited September 30, 2023. https://en.wikipedia.org/wiki/1986_FBI_Miami_shootout

Bushatz, Amy. "Shootout in Miami." Recoil Magazine. Accessed April 14, 2024. https://www.recoilweb.com/shootout-in-miami-142459.html

"Fatal Firefight in Miami." News. Stories. FB. April 22. 2011. https://www.fbi.gov/news/stories/fatal-firefight-in-miami

North Hollywood Shootout

Lloyd, Jonathan. "From the Archives: The 1997 North Hollywood Shootout." NBC 4, Los Angeles. February 28, 2023. https://www.nbclosangeles.com/news/local/from-the-archives-the-1997-north-hollywood-shootout/3103942/

"North Hollywood shootout." Wikipedia. Last edited April 11, 2024. https://en.wikipedia.org/wiki/North_Hollywood_shootout

Recoil Staff. "Anatomy of the North Hollywood Shootout: Tinseltown 2-11." Recoil Magazine. December 17, 2021. https://www.recoilweb.com/anatomy-of-the-north-hollywood-shootout-tinseltown-2-11-171988.html

Truman Assassination Attempt

"2 Shot in Seeming Effort to Kill Truman." United Press International. Accessed April 14, 2024. https://www.upi.com/Archives/1950/11/01/2-shot-in-seeming-effort-to-kill-Truman/7671541036971/ "Diagram of Assassination Attempt on President Truman at Blair House." Harry S. Truman Library/Museum. Accessed April 14, 2024. https://www.trumanlibrary.gov/photograph-records/77-3614

"Attempted Assassination of Harry S. Truman." Wikipedia. Last edited April 10, 2024. https://en.wikipedia.org/wiki/Attempted_assassination_of_Harry_S._Truman

History.com Editors. "An Assassination Attempt Threatens President Harry S. Truman." Updated October, 29, 2020. https://www.history.com/this-day-in-history/an-assassination-attempt-threatens-president-harry-s-truman

Hunter, Stephen and John Bainbridge. *American Gunfight: The Plot to Kill Harry Truman – and the Shoot-out that Stopped It.* New York: Simon & Schuster, 2005. Kindle.

Miscellaneous

"Action (firearms)." Wikipedia. Last edited November 10, 2023. https://en.wikipedia.org/wiki/Action_(firearms)

"Crime Data Law Enforcement Officers Killed in the Line of Duty Statistics for 2021." FBI. November 9, 2022. https://leb.fbi.gov/bulletin-highlights/additional-highlights/crime-data-law-enforcement-officers-killed-in-the-line-of-duty-statistics-for-2021#

"Distribution, Concentration and Rate of Fire. Fort Moore. US Army. Accessed April 14, 2024. https://www.moore.army.mil/Infantry/DoctrineSupplement/ATP3-21.8/appendix_f/CombatTechniquesofFire/DistributionConcentrationandRateofFire/index.html

"Door breaching." Last edited March 18, 2024. https://en.wikipedia.org/wiki/Door_breaching

"Firearms - Guides - Importation & Verification of Firearms - Gun Control Act Definition – Pistol." ATF. Updated April 27, 2018. https://www.atf.gov/firearms/firearms-guides-importation-verification-firearms-gun-Control-act-definition-pistol

Jones, Richard D. and Andrew White. *Jane's Guns Recognition Guide, Fifth Edition.* New York: Harper Collins, 2008.

"Vepr-12." Last edited June 11, 2023. https://en.wikipedia.org/wiki/Vepr-12

Manufacturers

Beretta.com

Colt.com

Glock.com

Heritagemfg.com

Hk-usa.com

Mossberg.com

Remington.com

Shadowsystemscorp.com

Sigsauer.com

Smith-wesson.com

Voere.com

Waltherarms.com

About the Author

Chris is a retired Green Beret with 26 years of service. He's deployed to Afghanistan, Bolivia, Honduras, Iraq, and a few other interesting places.

During his time with 20th Special Forces Group, he also worked in the Florida National Guard Counterdrug Program, where he provided training to law enforcement agencies.

In 2007, Chris got his first opportunity to share his firearms and tactics knowledge as a technical advisor for Scott Sigler's book *Contagious*. Seven years later, he spoke at ThrillerFest, then Bouchercon, and Quills.

He currently works as a contractor for the Florida National Guard's counterdrug school, MCTFT, where he writes classes about transnational organized crime and provides other curriculum support.

Chris moonlights as a technical advisor for writers. Head on over to TactiQuill.com if you're interested in his services.

He's written his first novel, though he's still trying to find it a home. You know how it goes.

If you enjoyed this book and found it helpful, please take a moment to leave a review on your favorite website.

You can also send questions and comments to Chris@chrisgrall.com. Who knows, your feedback could fuel a second edition.

www.ingramcontent.com/pod-product-compliance
Lightning Source LLC
Chambersburg PA
CBHW070052030426
42335CB00016B/1867